The Literature of Science

Edited by Murdo William McRae

The Literature

of Science

Perspectives on Popular

Scientific Writing

The University of Georgia Press   *Athens & London*

© 1993 by the University of Georgia Press
Athens, Georgia 30602
All rights reserved
Set in 10 on 13 Sabon by Tseng Information Systems, Inc.
Printed and bound by Thomson-Shore, Inc.
The paper in this book meets the guidelines for
permanence and durability of the Committee on
Production Guidelines for Book Longevity of the
Council on Library Resources.

Printed in the United States of America

97 96 95 94 93   C   5   4   3   2   1

Library of Congress Cataloging in Publication Data

The Literature of science : perspectives on popular
   scientific writing / edited by Murdo William McRae.
      p.   cm.
   Includes bibliographical references and index.
   ISBN 0-8203-1506-0
      1. Science news.   2. Scientific literature.   3. Communication
in science.   4. Technical writing.   I. McRae, Murdo William.
Q225.L58   1993
501′.4—dc20   92-22013

British Library Cataloging in Publication Data available

For Terri, Ian, and Carlin

Contents

Acknowledgments

Charles Anderson, "In Search of the Exact Location of the Soul: Richard Selzer and the Rhetoric of Surgery," from *Richard Selzer and the Rhetoric of Surgery*. © 1989 by the Board of Trustees, Southern Illinois University.

Andrew J. Angyal, "*The Immense Journey*: The Making of a Literary Naturalist," from *Loren Eiseley*. © 1983 by G. K. Hall and Company. Reprinted with permission of Twayne Publishers, an imprint of Macmillan Publishing Company.

Jeanne Fahnestock, "Accommodating Science: The Rhetorical Life of Scientific Facts," *Written Communication* 30(3): 275–95. © 1986 by Sage Publications. Reprinted by permission of Sage Publications, Inc.

Louis P. Masur, "Stephen Jay Gould's Vision of History," from *The Massachusetts Review* 30(3). © 1989 by The Massachusetts Review, Inc.

David S. Porush, "Making Chaos: Two Views of a New Science," *New England Review/Bread Loaf Quarterly* 12(4), Summer 1990.

Without the steadfast encouragement and support of all the contributors, who patiently endured my editorial proddings and pleadings, this book would never have been completed. I count myself particularly fortunate to have been able to work with them; even more, to learn from them. I am also grateful to Nancy Grayson Holmes, Kelly Caudle, and Melinda Conner, my editors at the University of Georgia Press, and to the anonymous readers of my manuscript. Whatever faults remain in this book are certainly not theirs. Finally, I am indebted to the Faculty Research Committee at Tennessee Technological University, which provided a reduced teaching load in order for me to work on this book.

Introduction: Science in Culture

MURDO WILLIAM MCRAE

Nearly thirty years ago, C. P. Snow wrote *The Two Cultures*, the seminal meditation on the relationship between the literary and the scientific cultures, to express his fear that the two might never bridge their "gap of mutual incomprehension" (1964, 4). The linguistic, literary, and theoretical perspectives that today guide cultural analysis make Snow's fear seem misplaced, however. These perspectives show us that science is situated in the culture that enables it, thus science should not be exalted over literature, history, philosophy, or other nonscientific cultural expressions. In the study of the nonscientific culture, critical relativism is more often the norm than not, with the consequence that few who study the nonscientific culture believe that multiple, partial, and conflicting perspectives involve the outright denial of truth.[1] Surprisingly, however, even those who grant relativism value for the study of nonscientific culture find it at times valueless for the study of science. For them, the conviction remains that science is not a relative matter but a matter of unimpeachable fact and truth.

But science *is* a relative matter, in the sense that *science must be related to the ideologies, values, habits of thought, and linguistic and rhetorical practices that shape our culture* if we are to understand both its power and its limitations. Finding this kind of relativism hard to accept, many traditionalists seek comfort in Francis Bacon's *Magna Instauratio* ([1620] 1937), often credited with being the philosophical origin of modern science. To read Bacon carefully, however, is to see that his arguments reveal something quite different from the traditional view that science escapes culture. Bacon thought that science must reign majestically over culture and that science was to have no master but nature, though even nature must finally submit to the power of science. Man may be "the servant and interpreter of Nature," and "Nature to be commanded must be obeyed" (272) proclaim two of the aphorisms by which Bacon hoped to supplant medieval science and institute the modern empirical method. Bacon further insisted that interpretive obedience to nature is the only legitimate route to "the most general axioms" (275) because such obedience permits "human knowledge and human power [to] meet in one" (272). But by making knowledge of nature synonymous with human power, Bacon contributed to the widespread belief that science might eventually master culture as it masters nature.

As his essentially political concern with human power illustrates, Bacon situated science in culture. This political emphasis also explains why Bacon knelt to the political ideology of his day when he dedicated his work to James I, submitting that "surely to the times of the wisest and most learned of kings belongs of right the regeneration and restoration of the sciences" (242). Perhaps the scientist must be dutiful to nature, but that duty must itself first be dutiful to political reality. Nor is this all. Just as he linked science to political authority, so Bacon hoped to draw authority from religion. This accounts for his doctrine of the "idols of the human mind," which prohibit us from understanding how nature reveals the "ideas of the divine" (275): science must never permit human nature (the idol of the tribe), individual bias (the idol of the cave), language (the idol of the marketplace), or specious philosophical systems (the idol of the theater) to impinge on its business. Bacon's doctrine is certainly a familiar one; even so, his reliance on the force of religion to insulate science from mundane human culture carries obvious political force. Only by rejecting all such false idols, Bacon claimed, could "the entrance into the *kingdom of man* . . . being not much other than the entrance into the kingdom of heaven" be assured (294; emphasis added).

Regardless of the familiarity of his argument that science must detach itself from the rest of culture, Bacon entangled science all the more forcefully in culture. So also does his most recent embodiment, Paisley Livingston, whose *Literary Knowledge: Humanistic Inquiry and the Philosophy of Science* (1988) takes issue with the view that cultural relativism has any value for the study of science. Indeed, Livingston scorns what he calls the "framework relativism" of recent literary and cultural theory, especially its conviction that language, habits of thought, and cultural values limit what scientists can think. The problem with that conviction, he claims, is that it distorts science, fostering the belief "that there can be no correct *epistemological* demarcation between science and nonscience" (1988, 23). Like Bacon, in other words, Livingston predicates his argument on something like a doctrine of the idols—in his case, literary and cultural theories that lead to the false belief that progress in scientific knowledge never occurs.

Livingston's stress on progress rests on a political agenda for science, however, thus indicating that his science is as culturally situated as Bacon's. Although Livingston writes somewhat critically of the separation between the sciences and the humanities in the university, he nonetheless insists that it would be naïve to question "the hegemonic scientific model," for to do so would be to ignore the enormous progress achieved by science (1990, 26). "Contradicted by widespread beliefs in the superiority of the scientific model of knowledge" (1990, 21), Livingston insists, cultural relativism in the humanities should defer to the truths that science reveals. Hence, in the same way that Bacon linked arguments for power over nature to the political authority of the monarchy and the church, so Livingston links arguments about progress in scientific knowledge to the lineaments of political power in the modern university. The players change, the argument remains: political power is licensed by what Bacon would call axiomatic, and Livingston hegemonic, knowledge of nature.[2]

I draw attention to Bacon and Livingston not to quibble with them but to underscore the conviction that informs *The Literature of Science*: we must be skeptical of any attempt to divorce science from the rest of culture. The essays in this book have all been written in the belief that science must be related to the values, ideologies, ways of thinking, and linguistic practices that shape our culture. Although considerable diversity in approaches and conclusions appears in what the contributors have to say about the literature of science, they all understand—whether they write about language and rhetoric; history, myth, and narrative; or culture and ideology—how inescapably science resides in the culture that enables it.

The first section, Language and Rhetoric, initially approaches the question of science in culture by taking up the relationship between the discourse that defines the scientist's professional role and the discourse he or she employs when writing for a popular audience. For Jeanne Fahnestock in "Accommodating Science: The Rhetorical Life of Scientific Facts," the relationship is problematic. Turning to the classical rhetorical tradition, she argues that popular science writing is overwhelmingly epideictic (concerned with praise), whereas professional science writing is dominantly forensic (concerned with fact). Emphasizing uniqueness and rarity, popular scientific discourse foregrounds the results of scientific research rather than the data on which they are based, often thereby removing the hedges that qualify professional scientific writing and making emphatic assertions which leave the popular impression that scientific truth is more certain than it actually is. Fahnestock also concludes that changes in information as it moves from professional to popular expression can be explained by classical stasis theory, the rhetorical tradition that defines the sorts of questions appropriate for a legal setting.

In "Popularization and the Challenge to Science-Centrism in the 1930s," Doug Russell offers a different account of what happens to professional discourse when it enters the public arena. Skeptical of strictly sociological and historical analyses of science popularization, Russell attends to the ways in which British and American scientists in the 1930s often espoused a science-centric belief in the superiority of science, even though their public statements about war, economic collapse, and political decay undercut that belief. Their popular expressions thus betray considerable "rhetorical strain," a mark of discursive opportunity, as Russell sees it, in the transition from the scientific world to the public. Not surprisingly, then, the 1930s witnessed an effective challenge to strict science-centrism in the hands of the mathematician Lancelot Hogben, whose popular science writing is distrustful of the "educated elite's linguistic exclusion of the common person from useful knowledge."

Continuing the discussion of the relations between professional and popular discourse, the next two essays examine how certain scientists have managed to communicate with the public in ways that overcome the restraints imposed by professional discourse. In "Loren Eiseley's *Immense Journey*: The Making of a Literary Naturalist," Andrew J. Angyal takes up Eiseley's effort to think beyond a materialist vision of evolution in order to reflect on the ways cultural evolution has relied on human language. Eiseley's meditations led him to the virtual invention of a new literary form, what he calls "the concealed essay." Extending Montaigne's personal and contemplative style and expanding "the personal essay for scientific purposes," Eiseley's essay form liberates him from the constraints of traditional scientific methods.

Like Eiseley, Richard Selzer also feels the constraints of scientific tradition; in his case, the tradition of the physician's seemingly necessary emotional detachment from his patient's suffering. Selzer was thus drawn to writing as a way of healing himself emotionally, writes Charles M. Anderson in his "In Search of the Exact Location of the Soul: Richard Selzer and the Rhetoric of Surgery." Powerfully imaginative and emotionally unsettling, Selzer's language seeks "to disturb and unsettle his readers" so that they "may move beyond the 'facts' of surgery toward its meaning." Selzer consequently eschews the vertical rhetoric of traditional medicine, which favors a hierarchical relationship between surgeon and patient. He favors instead a horizontal rhetoric of identification both informed by Emerson's conviction that the only true doctor is the poet and enriched by meditation on the ambiguous and multivalent meanings of words such as *wonderment,* *ecstasy,* and *grace.*

In "Oliver Sacks's Neurology of Identity," the final essay on language and rhetoric, I take up another physician's struggle with the constraints of a particular medical tradition. Sacks's sophisticated, even audacious, rhetoric appeals because it supports his argument for reforming the ways in which traditional neurology conceives the patient. Sacks maintains that the reductive and mechanistic neurological tradition fails to respect each patient's irreplicable individuality. Even so, Sacks's own Leibnizian and Pythagorean thought leads him to conceive the patient in equally reductive ways. For Sacks, as for the neurological tradition, each patient becomes a synecdoche for the patterns of disease and suffering replicated in every other patient. Sacks should not be faulted for bad faith, however. As we learn from what Michel Serres and Mikhail Bakhtin have to say about dialogue, empirical science requires the individual datum to be conceived synecdochically, as a replicable instance of the abstract patterns that formalize all the data.

The second section—History, Myth, and Narrative— departs from a primary focus on language and rhetoric, turning instead to how our culture's view of history, as well as our narrative and mythic practices, profoundly shape the literature of science. In "Stephen Jay Gould's Vision of History," Louis P. Masur observes that no other popular science writer has thought more deeply on these subjects than Gould. Contingency, not predictability, is the hallmark of history for Gould, the originator of the evolutionary theory of "punctuated equilibrium," which claims that long periods of evolutionary stability are disrupted by sudden, unexpected moments of change. Our culture's prevalent, essentially Darwinian, belief in gradual, predictable historical progress consequently troubles Gould, who is dedicated to the principles of democratic socialism, recognizing that too often a belief in predictability and progress has supported oppressive, even racist, social practices. Although he criticizes Darwinian notions of gradual evolution, Gould does not therefore subscribe to creationism, another challenge to Darwinian orthodoxy; as he has testified in recent court cases, creationism is nothing more than religion masquerading as science. Gould's scientific, political, and social views thus rest on his conviction that "scientists must recognize that ideas are culturally embedded," and that "they must appreciate the importance of narrative, of using words and images to tell stories that explicate change over time."

A concern with narrative practices is also central in the next two essays, in which Robert T. Kelley and David S. Porush examine several popular treatments of nonlinear dynamics and dissipative structures—scientific subjects often referred to as "chaos studies." In "Chaos out of Order: The Writerly Discourse of Semipopular Scientific Texts," Kelley examines how two of the important researchers in the area, Benoit Mandelbrot and Ilya Prigogine, translate their research for the popular audience. Linking Alan Gross's analysis of the rhetoric of overdescription in scientific texts to Roland Barthes's definition of "readerly" narratives, Kelley maintains that the effort to enclose meaning that operates in scientific texts fails because scientific texts are always polysemic, or, in Barthes's term, "writerly." Thus, even though Mandelbrot projects a readerly desire to be the progenitor of a new scientific field, his essay style, revisions of his major works, and debates with those who make popular use of his computer-generated graphics expose the writerly in his texts. By contrast, Prigogine (whose collaborator is the historian of science Isabelle Stengers) at first appears writerly when he insists that Newtonian science is not as closed as it once seemed. Even so, when Prigogine and Stengers claim that the study of irreversible processes heralds a new unification between science and culture, they betray their own readerly stance.

In "Making Chaos: Two Views of a New Science," Porush takes a different view of popular treatments of chaos studies, wondering why it is that Prigogine's work is completely disregarded in James Gleick's best-selling *Chaos: The Making of a New Science*. For Porush, Gleick's work is postmodern mythology in which the mythic hero is not a person but an idea, namely, Thomas Kuhn's notion that scientific change involves revolutionary paradigm shifts. But if this is so, why was Prigogine excluded from Gleick's pages, especially since Prigogine's Nobel Prize–winning work seems to reconcile evolution and the second law of thermodynamics? Prigogine told Porush that Gleick ignores Europeans, but Gleick challenged Porush to find a place for Prigogine in the story of chaos. Porush takes up that challenge, concluding that Gleick's narrative "is inspiring, exciting, dramatic, but philosophically impoverished" in its recuperation of a deterministic worldview, while Prigogine's work "entails a philosophical and discursive revolution" in the way it merges the human experience of time with the physics of irreversible systems.

Discussion of narrative practice in the literature of science continues with the next essay, Bruce Clarke's "Aspects of the Daemonic in Primo Levi's *Periodic Table*," Levi's semiautobiographical narrative of his life as a chemist. Levi invokes the mythic narratives of daemonic intermediaries such as Eros, Hermes, Proteus, Hyperion, and Apollo in his meditations on the metamorphic exchanges among sexual desire, commerce, theft, rhetoric, and science. These daemonic presences explain Levi's Oedipal desire as a chemist to overcome a feminized Matter in the pursuit of a masculine Spirit, a daemonic story further played out in his personal adolescent sexual crises. These presences also illuminate Levi's ambivalent struggle as a writer to identify himself with the daemonic, a struggle resolved in his final chapter, the tale of a carbon atom undergoing the complex transformations of photosynthesis and eventual metabolization in Levi's brain. Evocative of the quest romance, the story of carbon reveals that just "as the world of romance arises out of the anthropomorphic promotion of inert terms into animated figures, so life arises with the promotion of carbon from a chemical compound to a biological participant."

Paying attention to narrative and mythic practices is the business of hermeneutics, or the interpretation of "difficult linguistic or symbolic messages coming from remote or unfamiliar sources," as Martin Eger defines it in "Hermeneutics and the New Epic of Science." Eger takes exception with Gyorgy Markus and Jurgen Habermas, who insist that science has totally closed itself off from contact with everyday experience. Ranging across works by scientists as diverse as Jacques Monod, Douglas Hofstadter, Steven Weinberg, and E. O. Wilson, among others, Eger maintains that a new epic of science has emerged in the last twenty-five years. The epic is being told by visionary scientists convinced that there now exists "a truly seamless, thoroughly convincing, all-inclusive science of development," which brings unified scientific knowledge to bear on enduring cultural issues such as purposiveness and self-consciousness. If Markus and Habermas are right that scientific knowledge *as such* is no longer widely accessible, the new epic of science still functions as a hermeneutics of our culture's "cognitive ecology."

To examine the differences between professional and popular scientific language, or to show how science involves our culture's narrative and mythic practices, is also to raise questions about the ideological power that a Francis Bacon or a Paisley Livingston would confer on science. The refusal to grant validity to that conferral, as well as the sometimes disastrous consequences of accepting it, are the principal concerns of the last section, Ideology and Culture.

The section begins with Barry Pegg's "Nature and Nation in Popular Scientific Narratives of Polar Exploration." Pegg adopts a social-constructivist perspective when reading the popular scientific narratives of polar explorers such as Fridtjof Nansen, Roald Amundsen, Robert Peary, and Robert Scott. A careful reading of these narratives demonstrates how the failures of certain expeditions (Scott's, for example) can be traced to the explorers' "linguistic, institutional, commercial, geographical, ecological, and imperialistic" resistance to indigenous material nature and culture, as well as jingoistic condescension toward other explorers' nationalistic popular narratives, which often contained essential scientific and technological information. Even when, as in Peary's case, there was acceptance of indigenous culture, the Eurocentrism evident in his popular scientific writing still distorted his view of what Eskimo cultures had to offer. Only in Amundsen's popular narratives can we see a less than hostile or condescending stance toward indigenous culture, something approaching what Donna Haraway sees as essential for a reformation of the "traffic" between nature and culture.

Shifting to the contemporary scene and what may come to replace it, Alan G. Wasserstein's "Aggression and Power: The R-complex and Nuclear Blackmail" calls on what Michel Foucault has taught us about the relationship between power and knowledge. Wasserstein maintains that aggression names one kind of knowledge that scientific power creates in order to disguise itself. For popular writers such as J. D. Bernal, Arthur Koestler, Carl Sagan, and J. D. Franklin, Wasserstein shows us, aggression is known by virtue of its location in the primitive, reptilian midbrain, the R-complex. The task of science thus seems to be to control, even to eliminate, the savage effects of the R-complex, which now threaten us with nuclear annihilation. But when science proposes to overcome this threat through drugs, for example, or genetic engineering, Wasserstein perceives scientific and technological blackmail, not liberation. In the name of supplanting the supposed effects of aggression, of one kind of scientific knowledge, scientific power would merely create other kinds of knowledge, which would "paradoxically return us to a scene of magnified power." For Wasserstein, there is no universal liberation from scientific power; the best one can do is to ceaselessly interrogate it without succumbing to the delusion that it can be mastered: "Calling power relations into question does not abolish them; it realigns such relations, and hopes to do so without increasing them."

Although Mary Ellen Pitts shares Wasserstein's concerns about the power of scientific knowledge, she is more optimistic about the possibility of overcoming it. Her essay, "Reflective Scientists and the Critique of Mechanistic Metaphor," details how Eiseley, Sacks, Fritjof Capra, and the Russian polymath V. V. Nalimov are committed to overturning the metaphor, which dominates Western science. For Eiseley, mechanistic metaphor eradicates humankind's contact with nature, but contact can be restored through poetic knowledge, through projections of ourselves beyond the apparent boundary of our species. For Capra, the Cartesian dualism that underwrites the metaphor must be replaced by a Taoist awareness of the dynamic and nonlocal interconnectedness of things. For Sacks, reconceiving medicine as music and narrative can overcome the mechanistic reduction of the patient to a mathematical function. And for V. V. Nalimov, who fuses a sophisticated knowledge of the mathematics of probability with Sausserian perspectives on language, the metaphor can be displaced by visualizing the world not as a machine but "as a textual web of determinacies and indeterminacies."

Pitts's essay emphasizes the need to replace analytic tendencies in science, which separate and categorize, with synthesizing ones which value connection and unification. In "Contemporary Ecophilosophy in David Quammen's Popular Natural Histories," Allison Bulsterbaum Wallace examines how the same emphasis shapes Quammen's ecological thought. Skeptical of the anthropocentrism of Western thought and its theo-logic of human mastery over nature, Quammen's ecophilosophy insists on viewing "humankind as part of the biotic community, not a species removed from or superior to it." This skepticism, however, does not lead Quammen to advocate the Gaia hypothesis or to support the animal rights movement, ways in which anthropocentrism has otherwise been recently challenged. In fact, Quammen does not entirely relinquish anthropocentric thought, for in his ironic and self-reflective arguments he attempts to overcome anthropocentrism through individual acts of anthropocentric projection: by imagining how other creatures imagine us. In this respect, Quammen's work extends what Arne Naess calls "deep ecology," a way of seeing nature as a field, a set of relations, in which "each [creature] holds the other in special poise."

From the point of view of Baconian presumptions about the power of science and its detachment from the rest of culture, there is the tendency to insist that the popular culture simply cannot measure up to the superiority of scientific culture. In "*Omni* Meets Feynman: The Interaction Between

Popular and Scientific Cultures," however, David A. Stone concedes superiority to neither culture, arguing instead that a rapprochement between the two is not only possible but necessary. His comparison of texts by Nobel laureate Richard Feynman and *Omni* magazine writer Jessica Maxwell turns upon his insight that what fundamentally differentiates the two cultures is the way each organizes knowledge. "There is no 'natural' way to codify knowledge," Stone writes. "We learn to do so within our native culture." Hence there exists no superior way to organize what we know: popular culture relies on emotional response and a sense of community; scientific culture rests on a "schizophrenic split" between the abstract and the practical. True, popular culture defers to scientific authority, and scientific culture fosters its authority through a mythologized view of its own history. Even so, Stone concludes, the scientific and the popular cultures can speak to each other because both share traces of medieval culture. But science must jettison its mythologized history, recognize in its empiricism a link with the popular culture's valuation of the senses, and, in a spirit of scientific egalitarianism, reject the socioeconomic abuses that often attend scientific progress.

If there were somehow only one conclusion to be drawn from the essays collected in this book, it would be that science popularizations are anything but intellectually jejune. While it is true that science popularizations appear in a variety of sources, ranging from news-rack publications to university and learned society presses, it is precisely this range that should caution us about transporting the usual connotations of "popular" to a reading of what is often, and unwisely, characterized as the "popular scientific essay." That label unfortunately connotes what Richard Whitley rejects in the usual attitudes about science popularizations, namely, a demarcation between "a structured intellectual elite of knowledge producers" and a "diffuse mass of ignorant knowledge consumers" (1985, 6). That sort of distinction trivializes the reader of science popularizations as much as it demeans their authors.[3]

There are good reasons, then, for the term *literature of science*. By replicating the names of certain already well-established subdisciplines—history of science, philosophy of science, sociology of science—the term suggests a particular way of asking questions about the field of popular science writing. The term also indicates that the sorts of popular texts examined here are open to as full a range of contemporary interpretive techniques as any other works of literature. Finally, it emphasizes that the literature of science must be read not as mere popular transmission of superior scientific knowledge but as sophisticated production of knowledge in its own right.

The literature of science deserves respect equal to what is conventionally granted to professional scientific discourse. In one way or another, the contributors to this book grant that respect, even if they differ in their approaches and conclusions. But such differences, after all, are in keeping with the critical pluralism that has shaped this book. *The Literature of Science* in a sense, then, does return us to Bacon—not to his conviction that empirical science majestically commands all other knowledge, but to his desire to overcome the inertia of an ossified worldview: in his case, medieval Scholasticism; in ours, the monistic belief that science bears no relationship to our culture's values, ideologies, habits of thought, and linguistic practices.

NOTES

1. Several references at the end of this introduction list other approaches to the cultural embeddedness of science. See in particular Amrine (1989), Aronowitz (1988), Bazerman (1988), Bleier (1988), Foucault (1970), Gregory (1988), Haraway (1989), Latour (1987), Levine (1987), Lyotard (1984), Myers (1990), Peterfreund (1990), Shinn and Whitley (1985), and Woolgar (1988). Questions of the cultural embeddedness of science also involve a larger debate in the philosophy of science, between realism and representation. The subject is too broad to develop here, except to say that culture is a mediating category: reality constrains how our culture represents it, but cultural representation is all that we can say of reality; see Hayles (1990) for a discussion of this position. For additional discussion of realism and representation, see Feyerabend (1988), Kuhn (1970, 1977), Laudan (1990), Leplin (1984), Levine (1991), Livingston (1988, 1990), Longino (1990), Luntley (1988), Rorty (1979, 1982, 1985), Rouse (1987), and Van Frassen (1980).

2. Livingston presents another, somewhat more technical argument for scientific progress. He asks us to think of two knowledge states, k_1 and k_2, representing medical knowledge immediately before and after Harvey. For framework relativism, according to Livingston, k_2 is no closer than k_1 to K, or "the ideal of a perfect and total knowledge" of the human body. The meaning of scientific progress, Livingston continues, "resides not in the ideal as such but in the approximation of that goal, and it is in regard to this approximation that k_2 amounts to real progress over k_1, for k_2 is in fact a better approximation of the truth than was k_1" (1988, 28). There is at least one problem with Livingston's position, however. Since approximations presume a fixed point or known quantity (an approximation *to*, or *of*), they also presume complete, not ideal, knowledge of that which they less than

completely approximate. Livingston's idea of progress founders on a paradox, in other words: if one knows already what one approximates, how can one speak, as Livingston suggests, of not knowing it?

3. In *Bully for Brontosaurus*, S. J. Gould makes much the same point, contrasting the French attitude toward popular science writing (*vulgarisation*) with the American: the former "ranks within the highest traditions of humanism," while the latter "lies immured in deprecations" (1991, 11).

REFERENCES

Amrine, F., ed. 1989. *Literature and Science as Modes of Expression*. Dordrecht, Holland: Kluwer Academic Publishers.

Aronowitz, S. 1988. *Science as Power: Discourse and Ideology in Modern Society*. Minneapolis: Univ. of Minnesota Press.

Bacon, F. [1620] 1937. *Magna Instauratio*. In *Essays, Advancement of Learning, New Atlantis, and Other Pieces*. Ed. R. F. Jones. New York: Odyssey Press.

Bazerman, C. 1988. *Shaping Written Knowledge: The Genre and Activity of the Experimental Article in Science*. Madison: Univ. of Wisconsin Press.

Bleier, R., ed. 1988. *Feminist Approaches to Science*. New York: Pergamon Press.

Feyerabend, P. 1988. *Against Method*. Rev. ed. London: Verso.

Foucault, M. 1970. *The Order of Things: An Archaeology of the Human Sciences*. New York: Random House.

Gould, S. J. 1991. *Bully for Brontosaurus: Reflections in Natural History*. New York: W. W. Norton.

Gregory, B. 1988. *Inventing Reality: Physics as Language*. New York: John Wiley and Sons.

Haraway, D. 1989. *Primate Visions: Gender, Race, and Nature in the World of Modern Science*. New York: Routledge.

Hayles, N. K. 1990. *Chaos Bound: Orderly Disorder in Contemporary Science and Literature*. Ithaca: Cornell Univ. Press.

Kuhn, T. S. 1970. *The Structure of Scientific Revolutions*. 2d ed. Chicago: Univ. of Chicago Press.

———. 1977. *The Essential Tension: Selected Studies in Scientific Tradition and Change*. Chicago: Univ. of Chicago Press.

Latour, B. 1987. *Science in Action: How to Follow Scientists and Engineers Through Society*. Cambridge: Harvard Univ. Press.

Laudan, L. 1990. *Science and Relativism: Some Key Controversies in the Philosophy of Science*. Chicago: Univ. of Chicago Press.

Leplin, J., ed. 1984. *Scientific Realism*. Berkeley: Univ. of California Press.

Levine, G., ed. 1987. *One Culture: Essays in Science and Literature*. Madison: Univ. of Wisconsin Press.

———. 1991. Scientific Realism and Literary Representation. *Raritan* 10(4): 18–39.

Livingston, P. 1988. *Literary Knowledge: Humanistic Inquiry and the Philosophy of Science*. Ithaca: Cornell Univ. Press.

———. 1990. Literary Studies and the Sciences. *Modern Language Studies* 20(4): 15–31.

Longino, H. 1990. *Science as Social Knowledge: Values and Objectivity in Scientific Inquiry*. Princeton: Princeton Univ. Press.

Luntley, M. 1988. *Language, Logic and Experience: The Case for Anti-Realism*. La Salle, Ill.: Open Court.

Lyotard, J-F. 1984. *The Postmodern Condition: A Report on Knowledge*. Trans. G. Bennington and B. Massumi. Minneapolis: Univ. of Minnesota Press.

Myers, G. 1990. *Writing Biology: Texts in the Social Construction of Scientific Knowledge*. Madison: Univ. of Wisconsin Press.

Peterfreund, S. 1990. *Literature and Science: Theory and Practice*. Boston: Northeastern Univ. Press.

Rorty, R. 1979. *Philosophy and the Mirror of Nature*. Princeton: Princeton Univ. Press.

———. 1982. Method, Social Science, and Social Hope. In *Consequences of Pragmatism (Essays: 1972–80)*. Minneapolis: Univ. of Minnesota Press.

———. 1985. Texts and Lumps. *New Literary History* 16: 1–16.

Rouse, J. 1987. *Knowledge and Power: Toward a Political Philosophy of Science*. Ithaca: Cornell Univ. Press.

Shinn, T., and R. Whitley, eds. 1985. *Expository Science: Forms and Functions of Popularisation*. Dordrecht, Holland: D. Reidel.

Snow, C. P. 1964. *The Two Cultures: And a Second Look*. Cambridge: Cambridge Univ. Press.

Van Frassen, B. 1980. *The Scientific Image*. Oxford: Clarendon Press.

Whitley, R. 1985. Knowledge Producers and Knowledge Acquirers: Popularisation as a Relation Between Scientific Fields and Their Publics. In *Expository Science: Forms and Functions of Popularisation*. Ed. T. Shinn and R. Whitley. Dordrecht, Holland: D. Reidel.

Woolgar, S. 1988. *Science: The Very Idea*. Chichester: Ellis Horwood, Ltd.; London: Tavistock Publications.

Language and

Rhetoric

Accommodating Science: The Rhetorical

Life of Scientific Facts

JEANNE FAHNESTOCK

Whatever be the subject of a speech, therefore, in whatever art or branch of science, the orator, if he has made himself master of it, as of his client's cause, will speak on it better and more elegantly than even the very originator and author of it can.
—Cicero (1970)

Two thousand years ago, Crassus, speaking for Cicero in the dialogue *De Oratore*, could have been describing those prolific intermediates, the orators of magazine and newspaper columns who interpret the wonders of twentieth-century science for lay readers, accommodating new knowledge to old assumptions and trying to bridge the enormous gap between the public's right to know and the public's ability to understand. It is undoubtedly true that, with a few famous exceptions, the accommodators of science speak of it more elegantly than the scientists themselves. They communicate where the originators of new knowledge might only confuse. Nevertheless, a doubt is bound to occur: What happens to scientific information in the course of its adaptation to various noninitiated audiences? What, if any, changes does it undergo as it travels from one rhetorical situation to another? And how, in turn, is the discourse containing such information transformed?

In a foreword he wrote for Lincoln Barnett's popularization of the theory of relativity, Albert Einstein defined the Scylla and Charybdis of accommodated science writing:

Anyone who has ever tried to present a rather abstract scientific subject in a popular manner knows the great difficulties of such an attempt.

Either he succeeds in being intelligible by concealing the core of the problem and by offering to the reader only superficial aspects or vague allusions, thus deceiving the reader by arousing in him the deceptive illusion of comprehension; or else he gives an expert account of the problem, but in such a fashion that the untrained reader is unable to follow the exposition and becomes discouraged from reading any further. If these two categories are omitted from today's popular scientific literature, surprisingly little remains. (Barnett 1968, 9)

Einstein wrote this in 1948, and one cannot help wondering what his opinion would be now, after the explosion of scientific popularization that has taken place in the last fifteen years. To illustrate, among older established magazines, the circulation of *Science Digest* grew from 150,000 to 530,000, and *Scientific American* from 425,000 to 715,000, between 1970 and 1984. Over the same period, several new magazine-rack popularizations of science appeared, including *Discover, Technology Illustrated, Omni, Physics Today,* and *High Technology* (*Ulrich's International Periodicals Directory* 1984). The American Association for the Advancement of Science, the publishers of *Science*—which itself grew in circulation from 155,000 to 700,000 in the last fifteen years—sponsored what was for a time the most successful of these organs of accommodation at a circulation of almost 800,000: the magazine that changed its name every year, *Science79* to *Science85*. Established magazines like *National Geographic* and *Smithsonian* have also changed their editorial policies within the last few years to include more coverage of scientific subjects.[1]

At the same time there has been a wonderful proliferation of book-length translations of science—not all of them by Isaac Asimov. Many are written by practicing scientists who, like Carl Sagan, Stephen Jay Gould, James Trefil, and Lewis Thomas, have constructed a public voice. We have, for example, the prestigious *Scientific American Library*, in which scientists of the stature of geneticist Richard Lewontin, physicist Steven Weinberg, or chemist Peter Atkins explain, respectively, human diversity, subatomic particles, and the second law of thermodynamics. And we have series by accommodators like Jonathan Miller that aim a bit lower, *Darwin for Beginners* and *DNA for Beginners*; and even the "classic comics" of science, *The Cartoon Guide to Computers* and *The Cartoon Guide to Genetics*. One wonders how many of these avoid Einstein's double pitfalls of obfuscation and oversimplification.

Although the sociology of science and the corollary investigation of the rhetoric of scientific communication have expanded in recent years (Bazer-man 1984; Gilbert and Mulkay 1984; Latour and Woolgar 1979; Myers 1985), the study of the accommodation of science from expert to lay audiences is a relatively untouched subject. In my investigation of accommodated science writing from the rhetorician's perspective, I have located a number of paired communications that cover similar subjects but are addressed to audiences with different levels of background information and different degrees of interest. Much of my evidence comes from matched articles in *Science* and *Science82*, *Science83*, *Science84*, and *Science85*. Using this selected data, I make three interrelated observations: first, on the genre shift that occurs between the original presentation of a scientist's work and its popularization; second, on the change in "statement types" that occurs when a larger audience is addressed; and, third, on the usefulness of classical stasis theory in explaining what goes on in the "rhetorical life" of a scientific observation.

THE GENRE SHIFT

Aristotle's tripartite division of kinds of oratory provides a continually useful system for classifying discourse. Basically, Aristotle distinguished three types of persuasive speech—forensic, deliberative, and epideictic—according to purpose, audience, situation, and the time domain concerned. Forensic oratory is the oratory of the law courts, where litigants argue over the nature and cause of past events. Deliberative oratory has its place in legislative assemblies convened to debate the best possible course of future action. And epideictic oratory concerns a current, here-and-now judgment over whether something deserves praise or blame; funerals and awards ceremonies are the natural settings for epideictic discourse, which ultimately aims at solidifying the values of its audience (Aristotle 1984, 2159–61).

A case can be made for classifying original scientific reports as forensic discourse. Scientific papers are largely concerned with establishing the validity of the observations they report; thus the swollen prominence of the "Materials and Methods" and "Results" sections in the standard format of the scientific paper and the prominence given to tables, figures, and photographs that stand in as the best possible representation for the physical evidence the researcher has generated. Of course, scientific papers are also to some extent epideictic and deliberative; they cannot ignore creating a reason for their reporting. The point of making the reported observations has to be established in the opening paragraphs, and their place in an ongoing debate and the suggestions they yield about future work have to be established in the concluding "Discussion" section. But much of the relevance of scientific articles is extratextual—not spelled out in the discourse but supplied by context, by the assumed inferences the intended audience will make. One need only think of the deceptively simple statement with which Watson and Crick closed their initial paper on DNA: "It has not escaped our notice that the specific pairing we have postulated immediately suggests a possible copying mechanism for the genetic material" (cited in Judson 1979, 198). The original *Nature* audience immediately recognized the enormous consequences of the discovery; Watson and Crick could afford to be coy. In a similar way, scientific papers are, for the most part, explicitly devoted only to arguing for the occurrence of a past fact; significance is largely understood.

Accommodations of scientific reports, on the other hand, are not primarily forensic. With a significant change in rhetorical situation comes a change in genre, and instead of simply reporting facts for a different audience, scientific accommodations are overwhelmingly epideictic; their main purpose is to celebrate rather than to validate. And furthermore, they must usually be explicit in their claims about the value of the scientific discoveries they pass along. They cannot rely on the audience to recognize the significance of information. Thus the work of epideictic rhetoric in science journalism requires the adjustment of new information to an audience's already held values and assumptions.

Science accommodators who attempt to bring things down to the level of *National Geographic* or *Newsweek* or one of the science magazines have, at bottom, only two basic appeals to make in their epideictic arguments. For convenience I call these the "wonder" and the "application" appeals, corresponding to the deontological and teleological appeals in ethical argument. A deontological argument attempts to praise or excoriate something by attaching it to a category that has a recognized value for an audience. In science popularizations, all references to the amazing powers and secrets of nature or to the breakthroughs and accomplishments of the scientists themselves are basically deontological appeals. A teleological argument claims that something has value because it leads to further benefits. An epideictic argument praising the space shuttle, for example, would use the wonder appeal if it talked about the "never before" achievements of the machinery, astronauts, and engineers, and would use the application appeal if it pointed out spin-offs from the space program. If a scientific subject cannot be recast under these appeals, it is not likely to make its way to a wider audience. As a science writer for the National Institute of Dental Research put it, "Unless it's going to cost less or hurt less, the public doesn't want to hear about it."[2] Subjects in biology and medicine are naturals for these appeals, and so are disproportionately represented in science journalism. Subjects in mathematics, chemistry, and physics are much harder to accommodate.

THE CHANGES IN INFORMATION

Under the pressure of this genre shift from the forensic to the epideictic, it is not surprising that something happens to the information as it moves from one kind of discourse to another. To illustrate this change we can look at a simple example from a pair of articles, the first an original report of research that appeared in *Science* on the discovery of a carrion-eating bee, the second a short accommodated version of this article that appeared in *Science82*. Both magazines are published by the American Association for the Advancement of Science, but they are, of course, aimed at overwhelmingly different audiences—different in background information as well as purpose for reading. Accommodating the scholarly piece for the nonscholarly magazine is not, therefore, simply a matter of translating technical jargon into nontechnical equivalents. Though "mandible" becomes "jaw," "carrion" becomes "dead animals," and "masticate" becomes "chew" (interestingly enough, "regurgitate" stays "regurgitate"), the true accommodation involves finding the points of interest in the topic that will appeal to readers who are not apiologists or even specialists in any life science. (Some accommodation to a wider audience has gone on even in the original piece, which, after all, is not appearing in a journal devoted to bee experts.) In the different rhetorical setting, some of the "information" has changed. We can pinpoint some changes exactly by comparing sentences in the two versions. The original piece makes the following claim:

1a. No other protein sources are used by *T. hypogea* [the bee species under consideration], and pollen transporting structures have been lost, making this species an obligate necrophage. (Roubik 1982, 1059)

In the *Science82* version this becomes:

> 1b. Though other bees have teeth, this is the only species that cannot carry pollen. ("Vulture Bees" 1982, 6)

The change here is a subtle but significant one; the addition of "only" in 1b gives the second claim a greater degree of certainty than the first. The scientist who wrote the original report and who had just discovered a new species of tropical bee was not about to claim that no other similar species exist and that he had found the "only" one. Less cautious, the *Science82* writer has shifted this information a degree up in certainty. What prompts such a shift is undoubtedly the desire to add to the significance of the subject by claiming its uniqueness, its one-of-a-kind status. In the *Rhetoric*, Aristotle pointed out the perennial epideictic appeal that "a thing is greater when it is harder or rarer than other things" (1984, 2171).

The accommodated version also claims that the bees "eat any animal," an inferential extension from the diet observed and recorded in the *Science* article. This change is perhaps no more than an innocent hyperbole. But again it is an exaggeration in an interesting direction because it helps to glamorize the danger of the bees—if they eat any animal they could eat us—and glamorizing is the writer's purpose throughout the accommodation, part of the heavy task of bringing a deliberately dry research report into the realm of interesting journalism.

The claim of uniqueness serves the epideictic "wonder" appeal so well that we can find evidence of the science accommodation emphasizing the uniqueness of its subject, whereas the original *Science* report downplays it. The following paired statements come from articles about how cheetahs show amazingly similar blood profiles.

> 2a. The cheetah is unusual but not the only mammalian species with low levels of variation [in blood profiles]. The northern elephant seal (30), the moose (31), the polar bear (32), and the Yellowstone elk (33) have been reported to have diminished levels of variation. (O'Brien et al. 1983, 461)

> 2b. Such remarkably high levels of genetic uniformity are usually found only in specially bred laboratory mice. ("Copycat Cheetahs" 1983, 6)

The scientist-authors of 2a want to diminish the singularity of the phenomenon they have observed; because their purpose is to convince readers of the validity of their observation, the rarer the phenomenon, the harder their job. Their observations are more plausible if other, similar ones have been made, so they naturally cite analogous reports. But the science accommodator wants to make readers marvel at something, so this writer leaves out any mention of species that have shown similar genetic invariance and makes his subject seem more wonderful by claiming, in effect: "Here we have animals in nature exhibiting the genetic conformity of those bred for that very quality in the laboratory." The science accommodator is not telling an untruth; he simply selects only the information that serves his epideictic purpose.

The same pair of cheetah articles shows the tendency to exaggeration that also serves an epideictic purpose.

3a. The estimate [of genetic variety, or in this case, lack of it] is derived from two conventionally studied groups of genes: 47 allozyme (allelic isozyme) loci and 155 soluble proteins resolved by two-dimensional gel electrophoresis. . . . The entire cheetah sample was invariant at each of the 47 loci. (O'Brien et al. 1983, 460)

3b. But all the cheetahs carried *exactly the same form of every one* of the 47 enzymes. . . . In another test of *more than* 150 proteins, 97% of them matched in the cheetahs. ("Copycat Cheetahs" 1983, 6; emphasis added)

The original does not editorialize about the information it reports, but (relying on readers to do so) the *Science83* accommodation uses intensifying phrases—for example, "more than 150" when the total is precisely 155, and "exactly the same form of every one," a phrase that adds the ring of the carnival barker, whereas the original simply announces "invariant."

Along with claims of rarity and exaggerations, any assertion that something is "the first" of its kind is also a way to argue for its significance and value, as the following pair from articles on homosexuality demonstrates.

4a. This sex difference in the LH response to a neuroendocrine challenge is a critical feature in any evaluation of hormone responsiveness and sexual orientation: to our knowledge, this is the first simultaneous direct comparison of heterosexual and homosexual men with heterosexual women. (Gladue, Green, and Hellman 1984, 1497; second sentence, fifth paragraph)

4b. Some homosexual men have been shown for the first time to differ from heterosexual men in the way they respond to hormones. ("A

Biological Basis for Homosexuality?" 1984, 8; first sentence of the article)

In the accommodated article for lay readers, the claim that the study reported is the first of its kind is heightened by giving it the prominence of first-sentence position; the original mitigates this claim both by hedging it ("to our knowledge") and by burying it in the text.

Looking again at the articles on scavenger bees, we find another significant difference between an original report and its translation for lay readers. Based on his field observations, the scientist-writer who found the carrion-eating bees makes the following highly qualified claims:

> 5a. The bees masticate and consume flesh at the feeding site. They do not carry pieces of flesh to the nest, but *appear* to hydrolyze it with a secretion produced by either mandibular or salivary glands, which gives the feeding site a wet appearance. Individual bees captured while feeding, then forced to expel the contents of their crop were carrying a slurry of flesh, measuring between 37 and 65 percent dissolved solids by volume.[17] [Note 17 gives more precise information on how the collecting and testing were done, giving a "regress" of specificity for more inquisitive readers.] Bees tagged while foraging in the morning continued to depart and arrive at a carcass throughout the day, *suggesting* that animal food is passed by trophallaxis to other workers in the nest. Nest mates *may* then convert flesh into glandular substances. (Roubik 1982, 1060; emphasis added)

But in the accommodated version of essentially the same information, "appears" and "suggests" have vanished.

> 5b. The bees chew flesh after coating it with an enzyme that breaks it down. [The hydrolysis mentioned in the first version requires an enzyme.] They partially digest it, then fly back to the nest, where the substance is regurgitated to fellow worker bees. ("Vulture Bees" 1982, 6)

In the space limits of a short notice in a magazine of popularized science, there is no room for the qualifications a more knowledgeable audience would demand, qualifications that show the author's awareness of the criticism and refutation that an expert audience would raise against his inferences. To protect himself from such refutation, the scientist-author has naturally hedged his account. But because he fears no such challenge, the accommodator is far more certain of what is going on among the tropical bees. When qualifications are omitted, the result is greater certainty for the remaining claims. These omissions once again serve the accommodator's epideictic purpose, for only certainty can be the subject of panegyric. The public will be interested in these subjects only if they are significant, and there is simply no way to address the public with the significance of findings that are so carefully hedged that their reality seems questionable.[3]

Science accommodations also show another interesting tendency: they replace the signs or data of an original research report with the effects or results, once again increasing the significance and certainty of their subject matter. Scientists as authors retain wording as close as possible to their observed results, even though such a practice leads to complicated and verbose phrasing, whereas a popular account naturally replaces these substantives-as-signs with substantives-as-effects. In other words, accommodators leap to results, whereas the original authors stay on the safe side of the chasm. We can see this process going on in the following two excerpts from original and spin-off articles on the possibility of identifying a cancer genome.

6a. A similar analysis performed on the DNA taken from either the patient's normal bladder adjacent to the tumor, or from peripheral blood leukocytes, showed *the same two bands at 410 and 355 nucleotides,* indicating the presence of the same two alleles as were present in the patient's carcinoma. . . . Thus, *the alteration identified in this gene at the Nae I or Msp I site by restriction enzyme cleavage* appears to be in the germ line and must have existed before development of the bladder carcinoma. . . . Thus, it is tempting to speculate that there is an association between *this point mutation in the c-ras$_1$^{H}* gene and the bladder carcinoma. Although we have no information at present regarding the frequency of the mutant c-ras$_1$^{H} gene in bladder tumors, we do know that this change is infrequent in the general population since analysis of DNA from 34 individuals revealed the *presence of the Msp I/Hpa II site.* (Muschel et al. 1983, 855; emphasis added)

6b. Researchers from the National Cancer Institute and Yale University Medical School believe they have found, in both normal and diseased cells of a bladder cancer patient, a mutant gene that may have caused his malignancy. Their finding indicates that people

may inherit certain genes that predispose them to developing some
types of cancer. ("A Cancer Gene?" 1983, 10)

The noun phrases italicized in the first passage show the authors' tendency
to stay close to their precise experimental data; the bands were generated
by gel electrophoresis and the sites are positions where particular enzymes
have cut the DNA. These are signs. The popularization uses only the inferred artifact, the "mutant gene."

Science and *Science85* contain a pair of articles on bears that demonstrate still another telling difference between expert-to-expert communications and the overhearing that goes on in accommodation (Nelson, Beck,
and Steiger 1984; and "Hibernation: The Bear's Metabolic Magic" 1985).
Science popularizations not written by scientists themselves are not usually
based on published research alone; the compiling editors of science magazines also consult the original researchers in telephone or personal interviews. Thus the accommodated pieces often contain direct quotations from
the scientists in wording more straightforward than they are likely to use in
writing. In interviews the consulted scientists also make observations and
conclusions not found in the original articles aimed at peer audiences. Thus
the accommodated paragraphs on hibernating bears contain the principal
author's assertion, nowhere mentioned in the original research report, that
the bears bring the level of urea in their blood down by converting urea
into protein and in effect digesting it, a striking claim, given that it suggests
that a mammalian metabolic system has evolved the ability to turn a waste
product into food ("Hibernation: The Bear's Metabolic Magic" 1985, 13).
The *Science85* piece also claims that research into the metabolism of hibernating bears may someday "lead to substances that can promote similar
processes in humans with kidney ailments" who now depend on dialysis, a
very desirable spin-off from basic research indeed. It is easy to imagine the
prompting question from the science accommodator, who wants to elicit
a practical application, fulfilling the second of the two major appeals that
accommodated science articles can have. Because such speculative applications are rarely mentioned in reports to peers, they must be solicited off the
cuff. But are these speculations claims that the researchers could support
before a more critical readership? Or have they come "down" too quickly?

A slight legerdemain in phrasing, changing qualified claims into certainties and omitting contradictory evidence and giving space to unsupportable
claims, hardly seems of more than academic importance when the topics
are, among the articles sampled, bees and lizard tails and fly larvae and

hibernating bears and sailing clams and horses and jet lag. But what about subjects like the role of viruses in cancer or arthritis, the cholesterol factor in the diet, or the potential of recombinant DNA research? I have selected one topic in which the consequences of misunderstanding are far from benign and have followed it into popular accounts: the reported inferiority of girls to boys in mathematical ability.

In 1980, two Johns Hopkins psychologists, Camilla Benbow and Julian Stanley, reported in *Science* that seven years of screening for the mathematically precocious had netted far more boys than girls (Benbow and Stanley 1980). Because the researchers tested seventh-graders who presumably had all had the same academic exposure to math, the results weakened the hypothesis that a disparity in scoring was due to the fact that boys take more math courses than girls do. With astonishing rapidity, Benbow and Stanley's work found its way into *Newsweek*, *Time*, the *New York Times*, *Reader's Digest*, *People's Weekly*, *Science Digest*, *Ms.*, *Psychology Today* (and perhaps even the *National Enquirer*). These popularizations show the same tendency, observed above, to increase the certainty of the claims made in the original. There are other subtle and less than subtle differences created by titles, subtitles, artwork, omissions, and the juxtaposition of remaining points, as well as the changes of wording in the comparable statements that I focus on here. To give just a few of these other differences: the original article was entitled "Sex Differences in Mathematical Ability: Fact or Artifact?" a question that suggests the possibility of genuine debate or contradictory evidence. The title of a spin-off in *Time* magazine is "The Gender Factor in Math," a statement that presupposes the certainty of its referent in a way that a question does not. *Newsweek* at least keeps the question mark, but the title, "Do Males Have a Math Gene?" skews their coverage by suggesting that Benbow and Stanley observed a difference caused by inherent aptitude, not to mention the absurd suggestion that a single gene could be responsible for such a complex phenomenon as mathematical ability.[4]

The popularizations give some coverage to preexisting viewpoints that differ from Benbow and Stanley's, but this attention differs in the effect it can have, depending on whether the article ends with a disagreement or with a reiteration of Benbow and Stanley's position (or the popularizer's version of it). If Benbow and Stanley have the last word about anything, then it seems as if they have made a successful rebuttal of their opponents. In other words, although the newsweekly pieces may be following some journalistic principle of organization, inverted pyramid or "I" structure, they inevitably have argumentative structure and by their arrangement influence, even create, the reader's opinion.

Original and spin-offs also show great differences when we can match comparable statements. I have selected here the authors' concluding statements, which are carefully hedged in the original research report but appear much more certain when they are addressed to millions of readers.

7a. *We favor the hypothesis that* sex differences in achievement in and attitude toward mathematics result from superior male mathematical ability, which may in turn be related to greater male mathematical ability in spatial tasks. This male superiority is *probably* an expression of a combination of both endogenous and exogenous variables. *We recognize, however, that our data are consistent with numerous alternative hypotheses.* Nonetheless, the hypothesis of differential course-taking was not supported. *It also seems likely that* putting one's faith in boy-versus-girl socialization processes as the only permissible explanation of the sex difference in mathematics is premature. (Benbow and Stanley 1980, 1264)

The hedges and qualifications, which I have italicized in the quotation from the original research report above, disappear in the following popular accounts.

7b. The authors' conclusion: "Sex differences in achievement in and attitude toward mathematics result from superior male mathematical ability." ("Do Males Have a Math Gene?" 1980, 73)

7c. According to its authors, Doctoral Candidate Camilla Persson Benbow and Psychologist Julian C. Stanley of Johns Hopkins University, males inherently have more mathematical ability than females. ("The Gender Factor in Math" 1980, 57)

7d. Two psychologists said yesterday that boys are better than girls in mathematical reasoning, and they urged educators to accept the possibility that something more than social factors may be responsible. ("Are Boys Better at Math?" 1980, 107)

Newsweek in particular tended to sensationalize Benbow and Stanley's data. One of the researchers' samples, eighth-graders who took the test in 1976, was so small (only 22) that Benbow and Stanley explicitly omitted it when they reported the limits of their results: "To take the extreme (not including the 1976 eighth graders), among the 1972 eighth graders, 27.1% of the boys scored higher than 600, whereas not one of the girls did" (Benbow and Stanley 1980, 263). But *Newsweek*, searching for extremes to heighten the significance of its report, exercised no such restraint: "Among eighth-grade subjects in 1976, more than half the boys scored above 600 of a possible 800, but not one of the girls did" ("Do Males Have a Math Gene?" 1980, 73).

We could attempt to formalize observations of such changes in information between original and accommodated versions by borrowing the taxonomy of statement types suggested by sociologists Bruno Latour and Steve Woolgar (1979, 77–79) in their discussion of scientific discourse. Briefly, Latour and Woolgar distinguish among five types of statements according to the degree of certainty they convey. Type 5 statements are the most certain; they assert the sort of knowledge that seems self-evident to insiders, knowledge that surfaces only when an outsider's questions force the exposure of presupposed information. Type 4 statements consist of uncontroversial information that is nevertheless made explicit; scientific textbooks pass on the expressed certainties of type 4 statements, and accommodated science writing consists of type 4 and occasionally type 5 statements. The following sentences, for example, appear in a *Science84* notice of research on a possible arthritis virus.

> Simpson and his coworkers have now discovered that the agent, which they call RA-1, is similar to paroviruses—a family of viruses rarely found in humans. Paroviruses are extremely small, about a quarter the size of a flu virus; they have single-stranded DNA instead of double-stranded as do most DNA viruses; and they are usually resistant to heat and harsh solvents. ("Arthritis Virus" 1984, 8)

Latour and Woolgar would classify this definition of *paroviruses* as a series of factual type 4 statements. Because the anonymous author of this passage assumes some scientific knowledge on the part of the audience, type 5 statements do not appear. They would if a naïve reader asked for definitions of *virus* or *DNA*, or if a science writer, aiming for a wider audience, thought it was necessary to explain these basic terms.

Type 3 and type 2 statements have hedges, qualifications, or "modalities" that suggest that the information conveyed is not indisputable. In type 3 statements, the modalities can be subtle; just the citation of a reference or source following an assertion slightly weakens the certainty of a claim because it suggests the need for backing. That is the inverse of what we usually assume citations accomplish, but according to Latour and Woolgar's scheme, a quotation like the following is qualified simply because of the citation closing it: "The one example of a viral pathogen causing chronic arthritis of a mammalian host is the caprine arthritis-encephalitis retrovirus that elicits a proliferative synovitis and periarthritis in older goats" (Simpson et al. 1984, 1425). Citationlike hedges can also appear in the wording of a claim: "It was recently reported . . . that paroviruslike agents can be isolated from the synovial tissue of patients with severe RA disease" (Simpson et al. 1984, 1425). Type 2 statements are created when the qualifications are stronger; when, for instance, the wording draws attention to the availability of evidence or lack of it: "There is some evidence to support the notion that a series of events may be required for malignant transformation . . . and transformation of NIH 3T3 fibroblasts may represent only a subset of those events" (Muschel et al. 1983, 855). Such type 2 statements, prevalent in the examples like 6a taken from research reports, include words and phrases like *may, seems, suggests,* and *appears to be,* which convey the tentative status of the claim: "This result suggests that the mutant allele is present in the germ line of EK" (Muschel et al. 1983, 855).

Finally, type 1 statements are openly and frankly speculative, admitting the insufficiency of evidence and the very tenuous nature of a claim. Type 1 statements are most likely to occur in private discussions among scientists, but they occasionally appear in scientific papers: "Thus, it is tempting to speculate that there is an association between this point mutation in the c-ras_1^H gene and the bladder carcinoma" (Muschel et al. 1983, 855).

Latour and Woolgar's taxonomy attempts to be very sensitive to minute changes in the certainty of claims, and the changes demonstrated in the paired examples quoted above could be described as changes in statement types according to Latour and Woolgar's scale. Thus the change from 5a to 5b is a change from a type 2 qualification to a type 4 textbook-sounding certainty. In general, accommodated science writing traffics in statements of types 5 and 4, the exposed certainties, and of type 1, the weakly supported and speculative. Latour and Woolgar's scale may, however, introduce a specious rigor into the investigation of what happens to "information" as it travels from limited to larger audiences. After all, the degree of certainty conveyed by a statement may depend more on context than it does on wording. The hedges in Watson and Crick's notice were almost certainly not taken at face value by the original audience.

THE RELEVANCE OF STASIS THEORY

The pressure to be interesting is only one explanation of the changes in statement types and purpose that occur between scientific report and scientific popularization. Another explanation can be reconstructed, oddly enough, from *stasis theory,* a neglected component of classical rhetorical invention. Supposedly developed by second century B.C. rhetorician Hermagoras of Temnos in works now lost (Nadeau 1964, 370), stasis theory was fully explicated by Cicero in his *De Inventione* and *De Oratore,* by Quintilian in the *Institutio Oratoria,* and in the second century A.D. by Hermogenes in a very detailed treatise, *On Stases* (Fahnestock and Secor 1983, 136–37; 1985). Concerned primarily with legal argument, stasis theory defines and orders four kinds of questions that can be at issue in a criminal case: (1) What exactly happened and who did it? (2) What was the nature or definition of the act? (3) What is the quality of the act; or, in other words, what were the mitigating or aggravating circumstances? and (4) Who has jurisdiction in this case and what action is called for? Prosecution and defense tussle over the various issues, and if, for example, the defense loses or concedes on earlier accusations, it can take a stand at a higher level: "Well, yes, I did take the car on the night of the 18th, but it was really borrowing, not stealing"—a defense in the second stasis.

The practical system of ordered questions represented by stasis theory turns out to be a general scheme capable of accounting for the way issues naturally develop in public forums. People inevitably have to be convinced that a situation exists before they ask what caused it or move on to decisions about whether the situation is good or bad and what should be done about it and by whom. We can follow the stasis process with a hypothetical newspaper example: The news media inform us that a jetliner has been hijacked and a Middle East faction is responsible; we define this event as an act of terrorism, instantaneously judge it harshly, and debate over an appropriate response. In the West, and especially in the United States, there is a strong cultural presumption that any situation evaluated negatively (third stasis) demands reform (fourth stasis). So ingrained is this natural logic of issues in our debates that we inevitably move a topic through the four questions.

It is easy to see how stasis theory accounts for the changes in purpose and content between professional and public science reporting demonstrated above. An original, forensic scientific report engages an issue in the first, or conjectural, stasis: Does a thing exist? Did an event or effect really occur? Claims in the first stasis can be met with denials based on the evidence or on the definitions of key terms. In just this way Benbow and Stanley's first stasis report was bombarded with contradictions a few weeks after its appearance. For instance, the validity of using the Scholastic Aptitude Test as a test for mathematical ability (as opposed to achievement) was questioned. Notably, these formidable counterarguments appeared in *Science* after the public exposure of Benbow and Stanley's viewpoint, and these rebuttals were not subsequently reported in the popular press. Because they omit qualifications and contradictory evidence, accommodations like those made of the Benbow and Stanley report take it for granted that an issue is settled in the first stasis, and they move on quickly to the next stases: What is the reason for the effect? What value should be placed on it? and What, if anything, should be done about it? Thus the *Time* article affirmed male superiority in mathematical reasoning in its first sentence, in its second went on to ask why, and finally concluded by quoting the observation that girls should accept the difference and be helped to go on from there, corresponding, as it were, to the last stasis question calling for action.

The movement of a scientific observation through the stases, its "rhetorical life," is an inevitable consequence of changing the audience for a piece of information and thus the purpose of relating it and thus the genre of the discourse that conveys it. A wider, public audience is created by its concern with large public issues that affect the many—such as the mathematical education of boys and girls. And the audience I am talking about here is not so much a demographic classification as it is a reading role that anyone can adopt when reading a large-circulation publication. The *New York Times* or *Newsweek* audience would have no interest in staying dispassionately in the first stasis, unresolved between arguments over whether a certain observation is a fact or an artifact. Even if the scientific report were translated from insiders' to outsiders' language with the minimum amount of distortion and no attempt to provide an epideictic exigence for the report, the public as readers would move the information themselves into the higher stases and ask, "Why is this happening? Is this good news or bad news? What should we do about it?"

The way that information changes as a function of rhetorical situation certainly deserves scholarly scrutiny beyond this preliminary study, for at issue is the machinery and quality of social decision making in an expert-dominated age. The technique of analysis described in this article could be employed in any number of subject areas so long as the researcher finds similar subject matter being communicated to dissimilar audiences. Of particular interest would be publications that "translate" legal and financial information—new laws, procedures, entitlements—for the affected public. How, for example, does information about school lunch programs or about small business incentives reach its audience? Can information about a newly available service be separated from an epideictic framework that encourages or discourages an intended audience? Blandly stated information might be interpreted as institutional indifference or even as a warning to stay away, perhaps from an agency that disseminates public funds. The assumption held by some proponents of the "plain language" movement that meaning can be readily transferred from context to context by mere editorial wizardry needs a second and a third look (Siegel 1985, 98–99).

Another area the writing/rhetorical scholar should investigate is the use of scientific and technical information by political factions and lobbying groups. What happens to technical specifications when they levitate from the engineering manual or report to the briefing memo, the white paper, the money-generating mailing? In these cases, the context clearly switches from one that is fundamentally reportorial or archival to one that is frankly persuasive; the changes in content may be predictable. But what changes occur when the writer's purpose ostensibly remains constant through audience changes, when, for instance, a "second version" purports to be simply a summary or condensed form of the first? Selling summarized information is a growing business in the 1980s as the various segments of our society strive to keep up with one another. Scientists subscribe to abstract services, and state and local government officials pay to receive newsletters on a critical federal agency's latest policies. Print can seem too slow, and some of the latest information services come "on-line." Yet the above study of science reporting that condenses as it speaks to a different audience suggests that even abstracts and summaries may distort an original in critical ways.

Finally, the fundamental differences demonstrated above between writing for specialists and writing for different publics have certain pedagogical implications, particularly for writing-across-the-curriculum programs. Although the term *writing across the curriculum* and its acronym, WAC, have been pasted as labels on programs of considerable variety, the "purest manifestations" of the WAC approach, according to a recent review, are the "writing-intensive" or "writing emphasis" or "writing concentration" courses in which "students are taught to write by specialists within a particular discipline for the audiences and in the 'modes' and conventions of that discipline" (Griffin 1985, 402). The observations made in this study suggest that the kind of writing students are going to do in such courses will be of a very limited kind indeed; they will learn to write like specialists for specialists. Such writing components cannot replace a full, rhetorically based writing course for two reasons: first, they do not give students practice in addressing significantly different audiences and thus practicing the language skills that audience adjustment demands; and, second, they do not teach the public dimensions and responsibilities of specialist knowledge. The future engineer does not practice public accountability; the English major never tries to convince the uninitiated of the value of literary studies. Furthermore, WAC programs are fueled by certain pieties about "writing as a mode of learning"; they ignore the inevitable "addressed" or rhetorical nature of language and forget that one audience's learning is not another's. There is no "body of knowledge" without bodies of knowers, and these are multiple. A WAC program concerned with addressing only nonspecialist audiences would suffer from the same problem in reverse.

Ideally, students in advanced writing programs who are simultaneously taking courses in their specialties should have a full writing course that gives them extensive practice in addressing different audiences, specialist and nonspecialist, on subjects drawn from their majors. Only in such a course will students receive the kind of genuine writing instruction that makes "audience addressed" a reason for every language choice. And only in such a course will they experience the problems, moral as well as technical, of accommodating information for different genres, audiences, and purposes.

NOTES

1. Interview with Oliver Payne, writer, Cartographic Division, *National Geographic*, March 1985.

2. Interview with Pat Sheridan, National Institute of Dental Research, March 1985.

3. Another example of removing the hedges and qualifications comes once again from the articles on homosexuality cited in the text. The original report explicitly disclaims the causal connection between the hormone response studied and homosexuality, whereas the accommodation suggests the common assumption of a causal relationship between sexual orientation and genetic factors.

a. These findings are based on a particular subset of homosexual men and may not apply to all male homosexuals. Since we may have measured an adult hormonal correlate of sexual orientation that is causally independent of sexual differentiation, a causal relation should not be inferred. Unknown physiological factors in the adult may account for the differential responses of LH and testosterone reported here. However, even though a developmental relation between neuroendocrine response and sexual orientation is not certain, our findings are not inconsistent with such an interpretation. (Gladue et al. 1984, 1498)

b. Research with animals suggests that some differences between the sexes—females' tendency to be less violent, for instance—are shaped by hormones that begin affecting the brain even before birth. Gladue believes that biological factors may also predispose someone to be homosexual. ("A Biological Basis for Homosexuality?" 1984, 8)

4. In responding to letters of criticism in *Science*, Benbow and Stanley once again invoked the precise wording of their original conclusion: "So little of our report is quoted directly [in the letters] that it seems desirable to reproduce our concluding paragraph" (see 7a in the text). They were also aware of the misleading nature of popular accounts of their work: "We deeply regret that press coverage of our brief report confused the issues, rather than alerting people to the *magnitude* of the sex difference" (Benbow and Stanley 1981, 121).

REFERENCES

Are Boys Better at Math? 1980. *New York Times*, 7 December.
Aristotle. 1984. *The Complete Works*. 2 vols. Ed. J. Barnes. Princeton: Princeton Univ. Press.
Arthritis Virus. 1984. *Science84*, June.
Barnett, L. 1968. *The Universe and Dr. Einstein*. 2d rev. ed. New York: Bantam Books.
Bazerman, C. 1984. Modern Evolution of the Experimental Report in Physics: Spectroscopic Articles in *Physical Review*, 1893–1980. *Social Studies of Science* 14: 163–96.
Benbow, C. P., and J. C. Stanley. 1980. Sex Differences in Mathematical Ability: Fact or Artifact? *Science* 210: 1262–64.
———. 1981. Letters. *Science* 212: 118–19.
A Biological Basis for Homosexuality? 1984. *Science84*, December.
A Cancer Gene? 1983. *Science83*, April.
Cicero. 1970. *Cicero on Oratory and Orators* [*De Oratore* and *Brutus*]. Ed. J. S. Watson. Carbondale: Southern Illinois Univ. Press.
Copycat Cheetahs. 1983. *Science83*, October.

Do Males Have a Math Gene? 1980. *Newsweek*, 15 December.

Fahnestock, J., and M. Secor. 1983. Grounds for Argument: Stasis Theory and the Common and Special Topoi. In *Argument in Transition: Proceedings of the Third Summer Conference on Argumentation*. Ed. D. Zarefsky, M. O. Sillars, and J. Rhodes. Annandale, Va.: Speech Communication Association.

———. 1985. Toward a Modern Version of Stasis Theory. In *Oldspeak/Newspeak*. Ed. C. W. Kneupper. Arlington, Tex.: Rhetoric Society of America and National Council of Teachers of English.

The Gender Factor in Math. 1980. *Time*, 15 December.

Gilbert, G. N., and M. Mulkay. 1984. *Opening Pandora's Box: A Sociological Analysis of Scientists' Discourse*. Cambridge: Cambridge Univ. Press.

Gladue, B. A., R. Green, and R. E. Hellman. 1984. Neuroendocrine Response to Estrogen and Sexual Orientation. *Science* 225: 1496–99.

Griffin, C. W. 1985. Programs for Writing across the Curriculum: A Report. *College Composition and Communication* 36: 398–403.

Hibernation: The Bear's Metabolic Magic. 1985. *Science85*, January–February.

Judson, H. F. 1979. *The Eighth Day of Creation: Makers of the Revolution in Biology*. New York: Simon and Schuster.

Latour, B., and S. Woolgar. 1979. *Laboratory Life: The Social Construction of Scientific Facts*. Beverly Hills, Calif.: Sage.

Muschel, R., G. Khoury, P. Lebowitz, R. Koller, and R. Phar. 1983. The Human c-ras$_1$H Oncogene: A Mutation in Normal and Neoplastic Tissue from the Same Patient. *Science* 219: 853–56.

Myers, G. 1985. The Social Construction of Two Biologists' Proposals. *Written Communication* 3: 219–45.

Nadeau, R. 1964. Hermogenes' *On Stases*: A Translation with Introduction and Notes. *Speech Monographs* 31: 361–424.

Nelson, R. A., T. D. I. Beck, and D. L. Steiger. 1984. Ratio of Serum Urea to Serum Creatinine in Wild Black Bears. *Science* 226: 841–42.

O'Brien, S., D. E. Wildt, D. Goldman, C. R. Merril, and M. Bush. 1983. The Cheetah Is Depauperate in Genetic Variation. *Science* 221: 459–61.

Roubik, D. W. 1982. Obligate Necrophagy in a Social Bee. *Science* 217: 1059–60.

Siegel, A. 1985. The Plain English Revolution. In *Strategies for Business and Technical Writing*. 2d ed. Ed. K. Harty. San Diego: Harcourt Brace Jovanovich.

Simpson, R. W., L. McGinty, L. Simon, C. A. Smith, C. A. Godzeski, C. W. Boyd, and R. J. Boyd. 1984. Association of Paroviruses with Rheumatoid Arthritis of Humans. *Science* 223: 1425–28.

Ulrich's International Periodical Directory. 1984. New York: R. R. Bowker.

Vulture Bees. 1982. *Science82*, December.

Popularization and the Challenge to

Science-Centrism in the 1930s

DOUG RUSSELL

In Britain and the United States during the 1930s, concerned scientists engaged in an unprecedented public debate about the social relations of science. Their essays, speeches, radio broadcasts, pamphlets, educational texts, travel writing, and book-length studies carried the message, unsavory to many of their colleagues, that science is part of society and has social responsibilities. Among others, Gary Werskey (1978), William McGucken (1984), and Peter J. Kuznick (1987) have offered detailed historical accounts of the institutional and political manifestations and the theoretical and ideological intricacies of this debate. These accounts make clear that the widespread popularization of the debate was a key factor in its political success. Although Werskey's claim that the key British players in the debate "created from scratch an impressive literature on the social relations of science" (1978, 257) is difficult to accept, there is value, for literary and cultural studies, in examining the ways in which scientists navigated a difficult discursive terrain: the public self-interpretation of the social role of science. Questioning the extent to which scientists managed to move beyond the limits imposed by a science-centered worldview also provides an object lesson for those of us in language-centered disciplines seeking to engage in cultural critique.

To bring literary criticism into the interpretation of this significant political and cultural phenomenon automatically entails a concentration on the linguistic and rhetorical qualities of the texts produced, yet I contend that two factors affect the nature of such critical activity. First, any effort to assess the political or cultural meaning of these texts must not simply accept the context provided by sociological, historical, or biographical studies. Rather, it should recognize itself as an intervention in a present-day field of political and cultural dialogue (be it cooperative or contestatory), a field that should not be given the somewhat cozy label "interdisciplinary." Second, the study of so-called popular texts, with its implicit or explicit reference to the current burgeoning popularization of such issues as the green debate, raises the question of the development of an understanding of criticism as a player in the creation and maintenance of a sphere of public debate.

Drawing on recent thinking in the sociology and history of science, my literary critique of the "science-and-society" discourse of the 1930s focuses on how scientists used and interpreted the terms *science* and *society*. In doing so, I foreground the linguistic theory and practice of British biologist, socialist, and "scientific humanist" Lancelot Hogben, arguably the most successful popularizer of science of the 1930s. Most interesting is Hogben's interpretation of science as language and as discursive practice in literary and, I suggest, sociological and historical terms.

SCIENCE POPULARIZATION AND THE PUBLIC SPHERE: RHETORIC AS INTERPRETIVE TOOL

I take the term *public science* from a short piece by historian Frank M. Turner, who depicts science as inseparable from its social and cultural environment, and scientists as necessarily engaged in a dialectic of influence with state and social institutions, "upon whose good will, patronage, and cooperation they depend. The body of rhetoric, argument, and polemic produced in this process may be termed *public science*, and those who sustain the enterprise may be regarded as *public scientists*" (1980, 589). "Public science" falls within the broad topic of science popularization, the subject of recent study by sociologist of science Richard Whitley, who sees the popularization of science as more than an irrelevant add-on to the "knowledge production and validation process" (1985, 3). Scientists direct their popularizations to a wide range of audiences, including, among others, fellow scientists, funding bodies, and the general public. Popularization is thus the connecting link in a social system in which "reputational" organizations, the "primary social units of knowledge production and validation" (27, n. 22), participate in an intricate system of influence relations for which status and access to funds are the primary motivations.

In his survey of sociologists' changing views of popularization, Whitley comments that "the transformation of knowledge produced by one community into the language and concepts of another is very difficult, if not impossible, without seriously changing the nature of that knowledge" (7). Whitley tabulates varieties of popularization, differentiating between modes of expression on the basis of "degree of formalisation and technical precision" (16). Thus, he places the precise, formal, technical expression of communication among specialists at one end of the range, while at the other extreme is writing directed to a general public via the mass media, or to heterogeneous professional or advisory groups. The latter style presents science in a simplified form, "in everyday terms with vivid imagery and diffuse, discursive linguistic structures" (16–17). Similarly, Cloître and Shinn, in their extensive study of an expository continuum from "intra-specialist" to "being popular," on the basis of the parameters "argument, image and referent," construct a profile of popular science which in part classifies it as unsophisticatedly iconic and ambiguously metaphoric in imagery, as well as intuitive and subjective in argument (1985, 31–35).

These descriptions of the stylistic qualities of the popular science end of the scale indicate an intriguing mixture of style and effect. Whitley's observation that popularization may have dual- or multiaudience effectiveness suggests that popularization must be rhetorically multivalent to achieve broad appeal. Accordingly, to study popular science writing that aims to influence or persuade a mixed audience, we need literary interpretive tools that can account for a flexibility of language use. At the same time, however, literary criticism needs the specifying of social systems that sociology and history offer.

As a broad historical frame for my study, I find Jurgen Habermas's perspective on the decline of public debate particularly useful. With a certain nostalgia for the eighteenth century, Habermas writes of the emergence of the bourgeois public during the Enlightenment as a crucially important sphere for rational-critical debate, the means by which society challenged or kept in check the exercise of power by the state. According to Habermas, the disintegration of the public sphere since the late nineteenth century has paralleled trade protectionism, oligopolistic mergers, and other processes by which "social power became concentrated in private hands" (1989, 144). With "the transfer of tasks of public administration to enterprises, corporations, and semiofficial agencies" (151), "state and societal institutions fused into a single functional complex" (148). The most telling effect of this was a reduction of vital public debate to mere consumption of carefully packaged "debate" on radio and other mass media. As Habermas laments, "the web of public communication unravelled into acts of individuated reception" (161).

My analysis of 1930s public science writing operates within the historical model provided by Habermas and makes use of Whitley's view of the nature and function of science popularization. By focusing on the rhetorical complications of these texts, I underscore Turner's complaint that there is a problem in the too-literal, "uncritical reading of the documents of public science" (1980, 608) by historians. Historians and sociologists of science must attend more closely to the rhetorical and the linguistic.

TEXTS AND POLITICS: THE 1930S SCIENCE-AND-SOCIETY DEBATE

The writings of British physicist J. D. Bernal, a highly influential socialist theorist and activist, provide a good reference point from which to describe and assess the emergence of the science-and-society debate of the 1930s. In "The Scientist and the World Today," Bernal writes in a rhetorically peremptory fashion of "the possibility of immediate and practical action" (1933, 340) by scientists on the issues confronting society:

> By now, . . . scientists are visibly uneasy on the questions of war, unemployment and fascism, echoes of which have even reached the British Association. This uneasiness has to be crystallised in discussion and organisation.
> Individually scientists are weak and easily intimidated; once organised their key position in all modern states would make them a powerful force. (1933, 348–49)

Historians Roy and Kay MacLeod claim that the sense of unease, disorganization, and precariousness of position among scientists such as Bernal was largely prompted by the economic depression of 1929–34, which "ushered in a decade of violent economic and political change which played havoc with public, professional, and political attitudes toward science" (1976, 339). Other sociohistorical accounts of the politicization of science range from McGucken's exhaustively detailed empirical description of the institutional complexities of the social relations of science movement to Pnina Abir-Am's poststructuralist interpretation of a "diversified and pluralistic scientific order" (1985, 390) in the process of redefining scientificity and its own history and politics. My rhetorical analysis of science-and-society discourse goes further, however, by specifying the changes that occurred in the ways science presented itself to the general public. Initially, in accordance with Habermas's ideas, I will argue that the debate can be seen to operate on two levels: first, among a public of scientists; second, in the public sphere.

As implied by Bernal's reference to "even . . . the British Association," the initial response by scientists to external political issues and pressures was, at best, cautious and inward looking. Science was on the defensive. Using the established channels of communication within the scientific community, scientists set out in a rhetorically and linguistically defensive posture, hoping to promote science despite its association in the popular press with the horrors of chemical warfare and the increased destructiveness of weapons, and with technology-related unemployment. As such, the defense of science can be seen as debate operating initially in Habermas's "intraorganizational sphere"; that is, among scientists writing for a public consisting mainly of fellow scientists. What specialists wrote for each other shares two contradictory rhetorical characteristics: the negotiability of the meaning of the term *science*, and the maintenance of a firm hierarchical distinction between *science* and *society*.

To illustrate the negotiability of *science*, consider the economist Harold G. Moulton's introductory address on "Science and Society" at the 1938 Indianapolis conference of the American Association for the Advancement of Science (AAAS). Moulton raises the question "whether, as the years pass, science will prove a beneficent power continually advancing the welfare of the people who comprise society, or a social demogorgon" (174). Exploring this issue further, Moulton finds that "the term 'science' is in some ways ambiguous and confusing" (175). This confusion can be seen as a result not only of the interdisciplinary reworkings of what may be considered science, but also of a growing sense of the need to translate science to an ever-wider public. Indicative of this thoroughly decentered state of science, Moulton covers over ambiguity and confusion by reverting to a generalized abstraction, the "scientific spirit": "What we are really interested in here is the *scientific spirit*, which is an attitude of mind. . . . The objective, open-minded, scientific outlook need not of course be restricted to consideration of natural phenomena; it may and should pervade all other realms of investigation" (175).

An example of the rhetorical strategy of excluding society as "unscientific" is the address by the economist Josiah Stamp, president of the British

Association for the Advancement of Science (BAAS), to the Blackpool meeting held in 1936, the first dedicated to "Science and Social Welfare." *Science*, official journal of the AAAS, ran part of Stamp's address as its lead article only two days after the speech was made, an indication of the swell of interest in the issue and the international flavor it was taking. Imbued with the inveterate science-centrism that survived all the fluctuations of the science-and-society discourse in the 1930s, and perhaps later ones as well, Stamp's opening paragraph stresses that

> scientists see very clearly how, if politicians were more intelligent, if businessmen were more disinterested and had more social responsi- bility, if governments were more fearless, far-sighted and flexible, our knowledge could be more fully and quickly used to the great advan- tage of the standard of life and health—the long lag could be avoided, and we should work for social ends. (1936, 235)

Stamp vaguely defines the nature of science and its method in terms of what society apparently lacks—intelligence, disinterest, and flexibility. A vague entity qualified only by lack of science, society becomes an abstract backdrop to what is in essence a debate over the interdisciplinary politics of science. As his conclusion makes plain, Stamp's solution to the "duality which puts science and man's other activity in contrasted categories with disharmony to be resolved" (239) is simply to extend science into all facets of social life. Social problems "are but the fallings short of science" (239).

Though Moulton and Stamp represent the marginal science of eco- nomics, there is no significant difference between their science-centrism and that which emanated from the disciplinary strongholds of physics and chemistry. Take, for example, physicist Arthur H. Compton's speech, delivered in Ottawa in 1938 at the second AAAS "Science and Society" symposium, on the topic "Physics and the Future." Compton speaks of the growing influence of physical discoveries. He develops a picture of progress in communications technology effecting the integration of national and global society to the extent that "the world becomes almost a conscious unit, very similar to a living organism" (116; shades of the later "global village" metaphor). In presenting this historical world picture of the rapid growth of knowledge resulting in humanity "in a wholly new sense unified by science," Compton asserts that "science has thus become the basis of civilization," and, more specifically, that it is "physics . . . [that is] most significant in this regard" (116).

Another paper from the Ottawa conference, Harold C. Urey's "Chemistry and the Future," however, illustrates the nature of the science-and-society debate when it departs from the intraorganizational sphere and enters into the public. Urey's essay conveys both a conception of science as separate from society and a recognition, however limited, that science might be inseparably involved in, while being a major cause of, the "enormous social change which we [scientists] cannot see in the proper perspective because we are part of the phenomena" (1938, 135). He contrasts the scientific community's more ordered objectives with the "amazing" divergence in the objectives, constructive and destructive, of the whole community, yet depicts a situation in which science "must solicit the help of the community" (139). A tentative step toward theorizing the science-and-society issue, Urey's worldview is informed by the principle that the abundance of wealth and goods created by science must be distributed "to many people and not to the privileged few only" (139). Generalizing on this basis, he states: "Probably the business rivalries which led to the world war, communism in Russia, the invasion of Manchuria and China and of Ethiopia, and of world-wide depression, have all resulted from the great production of goods by scientific methods, together with archaic methods of distribution" (139).

All the texts mentioned above emerged within the traditionally, and assertively so, apolitical or moderate discursive environment of the British and American Associations for the Advancement of Science and their attendant journals, Nature and Science. The sense of uneasiness, which contradicts their inherent science-centric confidence, can be read not only in empirically economic, historical, and scientific terms, but also as a symptom of rhetorical strain accompanying the emergence of an altered (expanded, exploded?) discourse of self-interpretation by scientists from within a tightly constrained linguistic field.

SCIENCE GOES PUBLIC

Turning to the more truly public crystallizations of the science-and-society debate, let me now draw attention to "Science and Society," in which Bernal reviews the impact of the Soviet delegation to the Second International Congress of the History of Science and Technology in London in 1931. His remarks confirm my point that the rhetorical control of discourse proves difficult as the territory of commentary expands:

The Russians came in a phalanx uniformly armed with Marxian dialectic, but they met no ordered opposition, but instead an undisci-
plined host, unprepared and armed with ill-assorted individual philosophies. There was no defence, but the victory was unreal. The
strength of the spirit of bourgeois science, particularly in England, lies
in its avoidance of explicit statement. It is a comprehensive attitude
which cannot be effectively attacked because it is so genuinely implicit
and unconscious. (1931, 338)

At this point there is value in referring to Habermas's outline of the
conditions under which there may be some hope of a revitalization of the
public sphere. As he sees it, "under conditions of the large, democratic
social-welfare state the communicative interconnectedness of a *public* can
be brought about only in this way: through a critical publicity brought to
life within intraorganizational public spheres" (249). That is, Habermas
sees a necessary continuity between debate in the prepublic (intimate, private, intraorganizational) sphere and debate in the public sphere, a kind
of rehearsal taking place before making a public debut. In the practice of
the 1930s science-and-society debate, this continuity involved a move to
an even higher level of rhetorical and conceptual contradictoriness, characteristics at odds with the traditional notion of "rational-critical debate"
to which Habermas and the 1930s public scientists adhere. Whether this
is regarded as disabling or productive probably depends on one's sympathy with either a modernist or postmodernist perspective, a question I will
take up at the conclusion of this essay.

Also in the 1930s, an increasing organizational spread and fragmentation of the scientific community occurred, associated with the rise of
"outsider" scientific politics. In Britain and the United States, scientists dissatisfied with the political inertia of the established scientific bodies formed
a variety of interest- and issue-related pressure groups whose aims included
the creation of new structures within existing scientific associations and
government bureaucracies. Formed by scientists (and often, science journalists), these included the Association of Scientific Workers (AScW) in
Britain, America, Australia, and other countries; the American Committee
for Democracy and Intellectual Freedom; the Cambridge Scientists' Anti-
War Group; and the International Council of Scientific Unions. The independent Parliamentary Science Committee, established in Britain in 1933,
was, according to McGucken, the first social relations of science organization (1984, 5). A significant change wrought by such groups on mainstream

science was the formation, in 1938, of the BAAS Division for the Social and International Relations of Science.

Inspired by the Soviet Union's major promotion of science, these organizations directed considerable effort toward developing closer ties between science and government, including greater financial support. While the success of the bureaucratic institutionalization of the social relations of science movement perhaps led to a loss of its momentum and a decline in public debate, the attempt to move beyond the bounds of apolitical science entailed adopting a public stance from which to deliver a critique of scientists' apathy and disengagement and the alleged neglect of science by industry and government. This critique needed to take place either in journals devoted to a social or cultural criticism (such as the American Marxist journal *Science and Society* or the *Modern Quarterly* in Britain), in organization-sponsored publications (the AScW's *Scientific Worker*), in the products of sympathetic publishing houses, in public media open to the debate (the BBC and the science editorship of Waldemar Kaempffert at the *New York Times*, J. G. Crowther at the *Manchester Guardian*, and Ritchie Calder at the *Daily Herald*), or in public speeches. To varying degrees in all these media, the popularizers of the science-and-society debate theorized, or were conscious of, the linguistic demands of their public stance, and this awareness shows in their rhetorical strategies.

Take, for example, sociologist Read Bain, who, under the title "Scientist as Citizen," brands "pure" scientists as "bad citizens" who "fail miserably to discharge [their] civic duty" (1933, 414). Bain's argument for greater social and political engagement and advocacy on the part of scientists is thus a linguistically disruptive act:

> Have the learned social science societies, not to mention the still more smug biological and physical societies, ever taken a stand on such questions as the Harding Scandals, Mooney Case, Red Hysteria, Protective Tariff, Republican Prosperity, Harlan County Horrors, Gastonia Riots, Militaristic Madness, Labor Injunction Evil, Lynching Mania, Unemployment Disgrace? It would be "very unscientific," yet it is safe to say there is a high degree of agreement . . . on these and many other important questions of public policy. (414–15)

Although Bain calls for scientists to take a "stand" on public issues and reviles the apathy of "pure scientists," he grounds his case in the extravagant portrayal of scientists as "the inspired prophets and logical leaders of an age whose religion is science" (415). They must act, otherwise "Western culture is doomed to slow decay and final destruction" (415). There is no thought here that a social or cultural critique might require the modification of this grossly science-centric stance. Despite his contradictory assertion that, in one way, to be a pure scientist is to become antisocial, Bain nonetheless assures the reader that his "is not an attack on the scientific method as such" (413). Bain's text illustrates beyond doubt that in the public sphere the terms *science, scientific,* and *scientists* tend toward considerable rhetorical and conceptual ambivalence. I suggest that in feeding this tendency, Bain's text is involved in the initiation of discursive possibility; though Habermas might find its rhetoric antithetical to his ideal of rational-critical debate, a text such as Bain's can be described as publicity in its nascent stage.

Turning to science on radio, the question "What Is Science?" was the title of BBC adult education broadcasts by British physicist and socialist Hyman Levy, subsequently published in 1933. Levy sets out to thoroughly demystify science for his listeners, arguing the social roots of science and asserting that "science prides itself on [its] capacity for change" (1933, 34). Important for an interpretation of the function of the public science in which Levy was engaged is his version of the relations of science, politics, and industry. The familiar lament at the lack of vision on the part of "politicians and captains of industry" is directed toward the need for effective channels of communication and transference between "pure" science and industry, to bridge the "gap" between the "scientific man and the mechanic" (1933, 94–96). Levy tries to make a strong argument for "coordination" and "appropriate machinery . . . for application and development of scientific principles in industry" (1933, 91). However, as is the case in Bain's article, the lucidity and specificity of Levy's exposition mix with the science-centric bent of such general claims as "we have staked our future and the continued functioning of our civilized life on a complete belief in science" (1933, 87–88) and his concluding evocation of the "spirit of science" (1933, 106).

Levy's science-centrism indicates a swerve from his rhetorical enthusiasm for a progressive relationship between science and society. Even so, that enthusiasm returns in the introductory chapter of his 1938 volume *A Philosophy for a Modern Man,* in which he denounces the idealist notion of the scientific method, claiming it "breeds . . . a fictitious sense of detachment" (1938, 19). He argues that if the scientist's method "cannot be applied to the problems, undoubtedly serious, that arise from the social misuse of his work, for example the pounding to pieces of the town and inhabitants of Guernica, then his method has broken down when faced with a real situation of profound importance" (1938, 20). If the scientist is to become fully conscious of "his part in social practice," then, Levy claims, the scientist will need "to revolutionise the interpretation of his own activity" (1938, 20).

J. D. Bernal shared Levy's Marxian enthusiasm for the process of change. In "The Effect of Social Forms on Social Science," Bernal argues that if science, unaccustomed to any process of change more rapid than biological evolution, is to be applied "*consciously to human society, then we shall need to change science* and change it in a way as fundamental as that in which science changes society. We shall have to learn the active and not only the passive voice of nature" (1936, 114). Despite the rhetoric, however, Bernal's revolution in science turns out to be science's familiar reinterpretation of itself as the dominant player in the rough-and-tumble of social and political debate and action. This is why, as Bernal explains it, the outcome of science's association with industry has been, "in the course of the last century, imperceptibly to turn science into an institution, one comparable with and even more important than that of the Church or the Law" (1939, 11).

Bernal's stance as a critical "outsider" is complicated by the science-centric bent of his rhetoric; nonetheless, this can be seen as a healthy feature of writing that attempts to participate in the public sphere while maintaining the values of the private sphere. Recalling Habermas, we can relate this rhetorical mix to the mixed roles of the public scientist: as a participant in the private sphere of business (or the intraorganizational public sphere) and in the intimate sphere of the laboratory; and as a participant in the problematical discursive territory of the public sphere. Indeed, in an essay titled "Science and Liberty," Bernal points out the limited opportunities for scientists to participate in public debate: "On questions of fundamental social importance a few scientists, from the comparative security of academic posts, can and do speak out; the vast majority do not care or are afraid to say a word" (1938, 126). In a rhetoric characteristic of his Marxist views, he attributes this failing to the post–World War I professionalization of science, its "insidious disenfranchisement . . . in the capitalist democracies and its brutal subjection in fascist countries" (1938, 126), adding that "it has been directed up till now for private profit rather than public welfare" (1938, 128). Bernal's implicit and uncriticized ideal of a free and unfettered "science" shows, however, that it was as much their shared ideology of science-centrism as it was the political and institutional obstacles confronting popularizers that imposed limits on the extent of their public critique of the social relations of science. This point needs to be taken further, however, than historian Peter J. Kuznick is willing to go. Kuznick observes that "both the popular and scientific wisdom of the 1930s agreed that the catholicon of the decade was the 'scientific method' " (1987, 46). I maintain that closer reading of the rhetoric of popularizations of the science-and-society debate during that decade reveals the self-contradiction and fracturing of this ideology when it was exposed to the stresses of publicity.

TOWARD PUBLIC CRITICISM: HOGBEN'S LINGUISTIC TURN

By arguing that the practice of science is grounded in discourse, Lancelot Hogben advanced the social and cultural interpretation of science one very significant step further. The fact that his linguistic self-consciousness was founded on a strong theoretical base took Hogben's texts beyond a flair for well-chosen words, to a radical reframing of science in linguistic terms. Further, his achievement in putting language theory into practice is demonstrated by the enormous success his works had in reaching a large and varied audience. His two immensely successful "popular self-educators"—*Mathematics for the Million*, which topped the British best-seller list and by 1978 had sold half a million copies, and *Science for the Citizen*, which sold 200,000 copies (McGucken 1984, 93, n. 39)—are informed by his belief, stated elsewhere, that "freedom of discourse is a necessary precondition of socialism *en rapport* with scientific humanism" (1940, 22). These books thus represent a major stride in the move toward a popular writing that is inclusive rather than science-centrically (or intellectually) exclusive. In keeping with his decentered disciplinary and political stances, Hogben's social desideratum is "a less centralised, and therefore less bureaucratic and less congested, type of world organisation" (1940, 20). Accordingly, he points with concern to the rise of a new social class of skilled administrators, the "salariat," and warns that this influential group "may become the instrument for destroying democracy and freedom of discourse" (1940, 22).

For those of us imbued with the linguicentrism of literary studies, there is great appeal, of course, in the fact that Hogben's social critique gives priority, and a certain substantiality, to language. Where Bernal appears to prefer a bare, literal prose and has commented that he finds "most books to contain a disproportionate number of words to ideas" (1949, 339), Hogben's language makes much of its own opacity and rhetoricity. These qualities can be observed in his representation of the linguistic specialization and fragmentation of science: "The limitations imposed on scientific intercourse by the absence of any medium of common speech have multiplied, and friction between linguistic minorities now provides the sparks which may light a bonfire on which natural science is compelled to pour the petrol of its own destruction and that of civilization" (1940, 40). Here, Hogben employs an apocalyptic rhetoric to demonstrate that the failure of communication among specialists has destructive social consequences. The ironic spark of this rhetoric highlights both the theatricality of language and the knowledge that the written word must be used to effect change. By enlisting an extravagant metaphor to achieve his impact, Hogben is acting on his understanding of the importance of metaphor in science, a stance that has striking similarities with present-day literary analysis. That is, Hogben's self-conscious reinvigoration of the rhetoric of science stems from his claim that "the edifice of scientific knowledge is supported by a scaffolding of deceased and dying metaphors" (1940, 77). Although this comment refers to the language of science education, Hogben's attribution of social efficacy to language expands to include his insistence that efforts be made to create and maintain the "linguistic conditions of lasting world peace" (1940, 42).

Hogben's analysis of the practice of science popularization in the public sphere is also interesting in that the early theoretical basis of his linguistics was the principle of *publicity*. In the collection of essays entitled *The Nature of Living Matter*, he distinguishes between "a *public* world of common beliefs which all can share" and the "many *private* worlds which for the present remain impenetrable through the medium of discourse" (1930, 96). Hogben's *"publicist standpoint"* is an all-embracing one, taking philosophy and religion in its stride, and, as I read it, is a tool for understanding the workings of anything from specialist to popular discourse. The rise of science is presented not as a history of great inventions and inventors or institutions, but as primarily the development of "a new symbolism, the language of science," in which "not only sustained observation of nature, but active interference with nature, or experiment, is enlisted in the process of abstraction" (1930, 18).

As Hogben states, "the concept of *publicity* . . . is an essentially social one" (1930, 101), and *communicability* is its chief defining property. Most significantly, it offers a rethinking of science as an act of public communication: "The important feature about the world construction of science is not its externality but its communicability" (1930, 261). Publicity, grounded in linguistic communicability, represents a conflation of the social, material, and symbolic in the shape of "socialised reality" (1930, 262). This view has echoes in recent discourse theory; as such, it offers many of the attractions of what might be termed the ideology of linguicentrism while, paradoxically, working within a science-centric frame. This science-centrism is evident in Hogben's comment that "so soon as we engage in public discourse we are compelled to seek for a neutral ground. . . . This neutral ground is the public world of science" (1930, 261). Notably, however, within the constraints of this worldview Hogben has replaced the vague notions of the "scientific attitude" and the "spirit of science" employed by other scientists, offering instead a rigorous, pragmatic kind of objectivity, or rationality, which is achieved by submitting "to the discipline of discourse in practice" (1930, 246).

In a conceptualization that must have been partly influenced by the language theory being propounded at Cambridge by his contemporary Ludwig Wittgenstein, Hogben's scheme further allows for the practice of distinct "languages" within a culture. These embody different approaches to experience and constitute different grammars. Since these grammars are integral to material and social competence, experience and efficacy as a citizen are dependent on familiarity with more than one language. Hogben recognizes that languages are constantly changing. Thus his popularizing of science is spurred by the view that the comparatively rapid innovation in the language of science has denied public access and understanding, because the public is limited to the more conservative "ordinary language." Holding to "the possibility that science will in the course of time modify our habits of conversation in many directions" (1930, 313), he also accepts that science cannot thrive apart from common language.

In his later best-selling popularization *Mathematics for the Million* (1936), Hogben self-consciously fuses his linguistics with his democratic-socialist principles. In the introductory essay, "Mathematics, the Mirror of Civilization," he makes much of the history of the educated elite's linguistic exclusion of the common person from useful knowledge, and situates his popularization of science and mathematics in the context of the belief "that it is an essential part of the equipment of an intelligent citizen to understand this language" (1936, 27–28). Hence he advocates the role of "the democratisation of mathematics as a decisive step in the advance of civilization" (1936, 20). A knowledge of mathematics, "the grammar of size and order," is as necessary a survival skill in the face of "economic tyranny" (1936, 20) as knowing the grammar of ordinary language. To his concept of publicity, in other words, he adds the exercise of power in discourse: "the grammar of numbers was chained down to commercial uses before people could foresee the vast variety of ways in which it was about to invade man's social life" (1936, 21). Hogben's highly accessible commentary on "the cultural value of science" (1936, 34) is remarkable in that it brings together the claims of science and language theory. This can be read as expansionism on the part of the dominant ideology, science-centrism. However, such a view does not allow for the possibility that, in assuming a level of linguistic self-awareness, science has been significantly altered in Hogben's popularizations, in which science is viewed in its "interlocking with man's common culture, his inventions, his economic arrangements, his religious beliefs" (1936, 34).

My discussion of the popularization of the science-and-society issue by scientists has shown the rhetorical and conceptual contradictoriness that appeared when the ideology of "science is all" came under scrutiny. I have also suggested the limitations to a strictly historical or sociological analysis of science popularization. Even so, in order to engage in cultural critique, literary-critical analysis will need to modify its linguicentrism in various ways. First, a truly public criticism should be awake to the rhetorical self-contradictions that attend a mixing of in-house interdisciplinary debate together with debate in the public sphere. This will also, in the postmodern sense hinted at earlier in this essay, entail an acknowledgment of the rhetorical and conceptual contradictions within science popularization, and consequently a reworking of the idea of rational-critical debate. Second, the absence of a ready-made space for public critique will result in an uncomfortable disruption and dispersion of present critical practice, especially since the need to negotiate the study of popular science with other disciplines and interested groups will reshape theory and practice, whether we are conscious of it or not. Finally, the lack of women popularizers of science indicates the near exclusion of women scientists from the public sphere of debate on the science-and-society issue in the 1930s. Women scientists at Cambridge were proportionally far more politicized than the men; their participation in the debate at the intraorganizational level was significant; and Hogben and his scientist wife, Enid Charles, among others in the scientific community, were noted feminists.

That little or none of this crystallized into public debate highlights the need for interpretations of cultural and political phenomena to be alert to questions of gender in their theory and practice.

REFERENCES

Abir-Am, P. G. 1985. Recasting the Disciplinary Order in Science: A Deconstruction of the Rhetoric on "Biology and Physics" at Two International Congresses in 1931. *Humanity and Society* 9: 388–427.

Bain, R. 1933. Scientist as Citizen. *Social Forces* 11: 412–15.

Bernal, J. D. [1931]. Science and Society. In Bernal 1949.

———. [1933]. The Scientist and the World Today. In Bernal 1949.

———. [1936]. The Effect of Social Forms on Social Science. In Bernal 1949.

———. [1938]. Science and Liberty. In Bernal 1949.

———. 1939. *The Social Function of Science.* London: George Routledge and Sons.

———. 1949. *The Freedom of Necessity.* London: Routledge and Kegan Paul.

Cloître, M., and T. Shinn. 1985. Expository Practice: Social, Cognitive and Epistemological Linkage. In *Expository Science: Forms and Functions of Popularisation.* Ed. T. Shinn and R. Whitley. Dordrecht, Holland: D. Reidel.

Compton, A. H. 1938. Physics and the Future. *Science* 88: 115–21.

Habermas, J. 1989. *The Structural Transformation of the Public Sphere: An Inquiry into a Category of Bourgeois Society.* Trans. T. Burger. Cambridge: MIT Press.

Hogben, L. 1930. *The Nature of Living Matter.* London: Kegan Paul, Trench, Trubner.

———. 1936. *Mathematics for the Million: A Popular Self-Educator.* London: George Allen and Unwin.

———. 1938. *Science for the Citizen: A Self-Educator Based on the Social Background of Scientific Discovery.* London: George Allen and Unwin.

———. 1940. *Dangerous Thoughts.* New York: W. W. Norton.

Kuznick, P. J. 1987. *Beyond the Laboratory: Scientists as Political Activists in 1930s America.* Chicago: Univ. of Chicago Press.

Levy, H. 1933. What Is Science? In *Science in the Changing World.* Ed. M. Adams. London: George Allen and Unwin.

———. 1938. *A Philosophy for a Modern Man.* London: Victor Gollancz.

McGucken, W. 1984. *Scientists, Society, and State: The Social Relations of Science Movement in Great Britain 1931–1947.* Columbus: Ohio State Univ. Press.

MacLeod, R., and K. MacLeod. 1976. The Social Relations of Science and Technology 1914–1939. In *The Fontana Economic History of Europe: The Twentieth Century.* Part 1. Ed. C. M. Cipolla. Glasgow: Collins/Fontana.

Moulton, H. G. 1938. Science and Society. *Science* 87: 173–78.

Stamp, J. 1936. The Impact of Science upon Society. *Science* 84: 235–39.

Turner, F. M. 1980. Public Science in Britain, 1880–1919. *Isis* 71: 589–608.

Urey, H. C. 1938. Chemistry and the Future. *Science* 88: 133–39.

Werskey, G. 1978. *The Visible College*. London: Allen Lane.

Whitley, R. 1985. Knowledge Producers and Knowledge Acquirers: Popularisation as a Relation Between Scientific Fields and Their Publics. In *Expository Science: Forms and Functions of Popularisation*. Ed. T. Shinn and R. Whitley. Dordrecht, Holland: D. Reidel.

Loren Eiseley's *Immense Journey*:

The Making of a Literary Naturalist

ANDREW J. ANGYAL

"The most enormous extension of vision of which life is capable," writes Loren Eiseley in *The Immense Journey*, "is the projection of itself into other lives" (1957b, 46). This, his first book-length work, is a carefully orchestrated series of such extensions of vision, set within the controlling framework of evolution, that range forward and backward through time as they trace the emergence and development of life. Most of the thirteen chapters were originally published as separate essays and later rearranged to create a unified, imaginative account of how life evolved from its Precambrian origins to the magnificent complexity of man. The title of the book was inspired by a passage from Henri Frederic Amiel's *Journal Intime* (1883–84): "It is as though the humanity of our day, had, like the migratory birds, an immense journey to make across space" (Eiseley Collection, 1956 journal entry). Eiseley takes as his central theme speculations about humankind's recentness, physical development, and genetic endowments, and the enormous, interwoven complexity of life.

Although it was his first book to be published, *The Immense Journey* remains the touchstone for Eiseley's later accomplishments. Still the most popular of his books, it is also his work most accessible to the ordinary reader. Here the mature Eiseley voice first appears—scientifically informed, self-assured, and skeptical, yet still personal in tone and capable of wonder and imagination. A subdued romantic toughened by the rigors of his scientific training and by years of field research, he can still respond with awe to the marvels and mysteries of the natural world. His observations and meditations are records of a personal journey in which he seeks "to understand and enjoy the miracles of this world, both in and out of science" (1957b,12). Putting aside the once-strict empiricism of science,

Eiseley admits that he does not "pretend to set down, in Baconian terms, a true or even a consistent model of the universe" (1957b, 13). Instead, he proposes to offer "a bit of my personal universe" (1957b, 13), in terms as strange as those used by sixteenth-century geographic voyagers.

Eiseley conveys his musings with vivid, arresting images rather than abstract theories. His essays feature a variety of mammals, birds, and plants — fossil and living—set in dramatic landscapes of the past and present. They range through topics as diverse as the miracle of water, the appearance of flowering plants, the mechanisms of adaptations, and the mystery of the rapid emergence of the human brain. Several themes unify these speculations: an antimaterialistic bias, an opposition to scientism, and a desire to recover the past by using the imagination to transcend temporal restraints. Part of Eiseley's difficulty in organizing his account of man's "immense journey" lay in finding a narrative framework appropriate to such varied material as science, philosophy, poetry, and religion. Assistance from his editors enabled him to revise some of his published essays to form a continuous narrative within which he could explore the metaphor of evolutionary change as a series of extensions of vision.

Time is the dimension through which evolution unfolds, and Eiseley continually uses metaphorical comparisons to broaden our understanding of the magnitude of the earth's geological age. We are prisoners of time—of the present moment—far more than we are prisoners of physical location. Until we learn to perceive time and evolution in true perspective, we will not be able to see beyond the limits of a narrow anthropocentrism.

Evolution itself, the dominant theme of *The Immense Journey*, becomes "transcendental" in that all forms of life constantly strain against their limits, their ecological niches—in the quest for survival. Even the grotesque deep-sea fish "were all part of one of life's strangest qualities—its eternal dissatisfaction with what is, its persistent habit of reaching out into new environments and, by degrees, adapting itself to the most fantastic circumstances" (1957b, 37). Implied in these lesser extensions, moreover, is the ultimate mystery: man, who is himself, in Saint Augustine's words, "a great deep." All the important Darwinian concepts—inheritance, variability, competition, natural selection, divergence, and extinction—are present in *The Immense Journey*, together with more recent discoveries about human evolution. Evolution becomes less a hypothesis, however, than a controlling metaphor to Eiseley, itself capable of manifold extensions and applications in the mind of a poet-scientist. To accept the idea of evolution is to accept a world of infinite possibility. Thus, for instance, in "The Flow of the River," an "evolutionary leap" is embodied in the actual leap of a channel catfish Eiseley had rescued from the frozen Platte River and kept for the rest of the winter in a fish tank, only to find the creature dead one morning on the basement floor (1957b, 23–34).

The Immense Journey divides roughly into three major thematic sections: life before man (chapters 1–5), the emergence of man and his future prospects (chapters 6–10), and the mystery of life (chapters 11–13). In the first section, each chapter introduces at least one key evolutionary concept, beginning with Eiseley's discussion of evolutionary time in "The Slit," the significance of water as the medium of life in "The Flow of the River," the oceanic origins of life in "The Great Deeps," and the adaptability of fauna and flora in "The Snout" and "How Flowers Changed the World."

Chapters 4 and 5 introduce the mechanisms of evolutionary change. Eiseley shows how in each successive age the dominant forms arose from simpler animals that could adapt because they were not restricted to a specific environment. Overspecialization often leads to extinction of species when the conditions of life suddenly change. In the myth of the "snout," or Crossopterygian (a primitive fish), Eiseley recounts the epic story of how animals first came ashore during the Devonian period. Thus he illustrates the concept of evolutionary succession (Edward Drinker Cope's "law of the unspecialized") with a vivid and imaginative reconstruction of a prehistoric event (1957b, 55).

The world we accept as "natural" today is actually the result of a series of complex adaptive responses made possible by the emergence of the angiosperms (flowering plants) less than a hundred million years ago. This new class of flora, with its nectar and wind-carried or insect-borne pollen, its fertilization within the flower, its fruits, its seed cases, and its elaborate transport mechanisms for those encased seeds—which Eiseley calls "capsules of life" (1957b, 69)—made possible a series of parallel faunal adaptations. With nectar and pollen providing a new food source for them, insect varieties multiplied. The early mammals—Eiseley's "shabby little Paleocene rat" (1957b, 8)—now had the food necessary to maintain their interior body temperatures at a consistently high level, which permitted the evolution of internal temperature regulation and correspondingly high metabolic rates. The fruits and seeds would also provide food for another evolving group of animals—the feathered lizards, or birds. Thus angiosperms, mammals, insects, and birds all coevolved, as each in complex ways made the others' evolutionary success possible.

The next five chapters of *The Immense Journey*, beginning with "The Real Secret of Piltdown," deal with the emergence of man and his future prospects, an issue of particular interest to Eiseley, a physical anthropologist. Primate evolution was made possible by a series of climatic and ecological changes early in the Cenozoic era, some seventy million years ago. These changes did not ensure man's appearance or survival as a species, however. How man managed to emerge so quickly in evolutionary terms and to dominate his world so completely has been a matter of dispute among scientists for more than a century. Charles Darwin and Alfred Russel Wallace, for instance, offered quite different theories to account for man's success. Darwin argued for the slow and gradual evolution of *Homo sapiens* and believed that human evolution could be explained largely in terms of competition and natural selection. Wallace, on the other hand, believed that man had emerged quite recently as a species. He saw unexplained forces at work in shaping human evolution, particularly in the rapid increase in the size of the human brain.

Although recently discovered hominoid fossils support Darwin's gradualist position and argue for a longer period of human development, Eiseley still rejected a strictly materialistic explanation of human evolution. He never accepted the implicit mind-matter dualism of science, which left an unaccountable gap between matter and consciousness. There were too many unresolved questions, he believed, to trace the development of human consciousness entirely through mechanistic causes. Nor could he satisfy himself that the "creativity" of natural selection alone could explain man's uniqueness and cultural achievements. Instead, Eiseley and Wallace both emphasized cultural rather than biological evolution in explaining man's distinctiveness.

These contrasting views of evolution have profoundly different implications in terms of human nature. A "pure" Darwinism seemed, to the Victorians at least, to sanction human aggression as justifiable or even beneficial from an evolutionary perspective. Darwin's critics argued that natural selection condemned humans to endless struggle, with the strong and ruthless surviving. This view of human nature has been used to rationalize all manner of oppression and injustice and to minimize humans' artistic ability and altruistic impulses. Darwin insisted that these cultural accomplishments could be explained in biological terms, whereas Wallace argued that natural selection and the struggle for existence could not alone have produced these artistic, mathematical, and musical abilities. Nor did man's increased brain size automatically make him civilized, since his brain capacity probably increased before he was fully able to use it: if indeed he can fully use it yet.

Much of the problem for anthropologists arises from the lack of a reliable fossil record for early man. Unlike the remarkably complete paleontological evidence for the evolution of the horse, the evidence of human evolution is at best fragmentary and incomplete. Entire theories have been based on a few disarticulated bones. With so little information available it seems premature to speculate about the physical development of *Homo sapiens*. Eiseley particularly opposed the reductionistic implications of a strict Darwinism that would deny our intangible but uniquely human qualities. As he demonstrates in his conclusion to "The Real Secret of Piltdown," the most important result of the discovery of this hoax was that it forced scientists to reconsider their assumptions about the development of the human brain.

In the next chapter, Eiseley compares our evolutionary past to a "maze," a confusing network of descent from our tarsioid or anthropoid ancestors with no clear evolutionary pattern. Natural history is a labyrinth from which we draw subjective impressions of human nature. Like the witches in *Macbeth*, it confirms what we wish to see in ourselves. This confusion is reflected in the debate among scientists about human origins. Some imagine that we are descended from a "homunculus," or "little man," while others regard the ancestor as a "shaggy anthropoid" or "ape man." Eiseley finds insufficient evidence for a clear verdict either way. As he notes in an unused epigraph taken from Henri Bergson's *Creative Evolution* (1911), "the route we pursue in time is strewn with the remains of all that we began to be, of all that we might have become" (109). What interests Eiseley most is the combination of physiological and anatomical adaptations that mark human evolution: bipedal locomotion, prehensile hands, color vision, stereoscopic (front-set) eyes, prolonged infancy, and expanded brain size. These physical changes, in turn, made advances in human culture possible, such as the use of tools and weapons, the discovery of fire, and the development of agriculture through the domestication of certain plants and animals.

Man is what Eiseley calls a "dream animal" (1957b, 120) because he was able to use the evolutionary advantage of his enlarged brain to develop symbolic communication—language—and through words to attain self-consciousness and memory. With man, life had at last become aware of itself. In a hauntingly beautiful passage, Eiseley describes this mythic moment of man's departure from the "Eden" of the instinctual world to the awesome knowledge of self: "For the first time in four billion years a living creature had contemplated himself and heard with a sudden unaccountable loneliness, the whisper of the wind in the night reeds. Perhaps he knew, there in the grass by the chill waters, that he had before him an immense journey" (1957b, 125–26). The story of Eden becomes here an allegorical account of man's passage from the certainty of instinct to the vast, uncertain world of knowledge, self-consciousness, and choice.

In "Man of the Future," Eiseley dismisses predictions about man's future appearance by demonstrating that the hypothesized features would not be new. An extinct line of early African anthropoids, "Boscop Man," possessed these same "ultramodern" features—large cranium and skull volume, high forehead, small teeth, and delicate, childlike features. Yet the Boscop people disappeared; perhaps because they were unable to compete with aggressive neighbors or, more likely, because they did not develop a sustaining culture. Their large brains, larger than those of modern humans, apparently went unused. Again Eiseley stresses the importance of cultural rather than physical features in defining human nature.

"Little Men and Flying Saucers" continues Eiseley's discussion of man's physical and cultural uniqueness. Man was neither "prefigured" in earlier forms of life or in the pattern of creation, nor is he likely to find his counterpart anywhere else in the universe. He is the solitary and particular creation of a set of biological conditions unlikely ever to be duplicated. To believe otherwise is to indulge in the conceit of a human-centered universe, whether we believe that previous forms of life point toward humans or that the human drama unfolds on other worlds. Man's prospects, according to Eiseley, are not buried in the past or hidden in some obscure future, but lie within himself, latent, in the dreams and visions he projects onto the world.

Since the narrative perspective of *The Immense Journey* continually moves from the specific toward the abstract, it is appropriate that the final three chapters—the section on the mystery of life—contain a philosophical summation. Here Eiseley offers what might almost be called a "vitalistic" affirmation of life. In each of these chapters he asserts opposition to scientistic assumptions that account for life entirely in chemical or physical terms, or else view living creatures as nothing more than complex, sentient machines. Through a series of visionary moments Eiseley shows how repugnant such a philosophy was to him. The soaring of pigeons above a city street, the reanimation of inert chemicals in a flight of warblers over the badlands, the cry of the song sparrows against "Cain" the raven, the heroism of the spider, and the reunion of the hawks are all eloquent replies to the reductionism of science. For some readers these passages might seem sentimental, but the emotional conviction in these anecdotes always points to some broader, though unstated, meaning. It is never cultivated for its own sake. Evolution becomes such a potent metaphor in Eiseley's work because through it he implies that there is something mysterious, purposeful, even transcendent about these apparently random events in the natural world.

In "The Secret of Life," Eiseley suggests that, having displaced the biblical creation myth, contemporary science is obliged to create a mythology of its own to answer the ultimate questions. Though insisting upon the inscrutability of life, he hints that its "secret" may be contained in the ability of living organisms to reproduce their precise pattern, structure, and form through genetic codes. For him, the living creature is always greater than the sum of its parts. Reductive analysis and an "uncompromising materialism" will not yield the secret of the creation of even the simplest one-celled organisms from their constituent molecules.

Clearly a romantic, and perhaps even a vitalist at this point, he hopes that the secret of life will continue to elude the scientists in the laboratory. Eiseley then compares the mystery of organic form to a "dance of the molecules," a metaphor that calls to mind the graceful patterns and configurations of living structure, as well as the dominant Renaissance symbol for the harmony of the universe, found in Sir John Davies's poem "Orchestra." His appreciation of the patterns of nature is finally more aesthetic and religious than analytic, so that *The Immense Journey* ends on a note of unorthodox science, with Eiseley's affirmation of the potential for life contained in all matter:

> Rather, I would say that if "dead" matter has reared up this curious landscape of fiddling crickets, song sparrows, and wondering men, it must be plain even to the most devoted materialist that the matter of which he speaks contains amazing, if not dreadful powers, and may not impossibly be, as Hardy has suggested, "but one mask of many worn by the Great Face behind." (1957b, 210)

THE VICTORIAN NATURALIST

Dissatisfied with the restrictive, value-free orientation of modern science, Eiseley turned to the essay in the hope of finding a form through which he could articulate his sense of wonder. For some time he had been privately disillusioned with the "religion of science" (Eiseley Collection, undated entry)—with its rigid assumption that every natural event in the universe can be rationally explained by prior events. He found it increasingly difficult to reconcile his "personal universe" of mystery and beauty with the rational universe of science (1957b, 13), where everything is ultimately reducible to fact and measurement. No longer did he wish to maintain the scientific detachment that would not permit an imaginative or aesthetic response to his study of early man. "I no longer believe that science will save the world," he commented in one of his notebooks (Eiseley Collection, undated entry).

"Anthropology," Eiseley once wrote, "is the science of man eternally trying to understand himself and never succeeding" (Eiseley Collection, undated entry). He envisioned anthropology as a human science, admitting a variety of styles and approaches, yet he found himself dismissed by his colleagues as a "popularizer" or a writer of "inspirational literature."[1] Faced with this kind of determined hostility to his imaginative style, Eiseley turned from conventional science to a form of personal expression more compatible with his sensibility, that of the literary naturalist. In "The Judgment of the Birds," he describes his purpose as a seeker of "natural revelations" for a culture that has lost its sense of wonder:

> It is a commonplace of all religious thought, even the most primitive, that the man seeking visions and insight must go apart from his fellows and live for a time in the wilderness. If he is of the proper sort, he will return with a message. It may not be a message from the god he set out to seek, but even if he has failed in that particular, he will have had a vision or seen a marvel, and these are always worth listening to and thinking about. (1957b, 163)

These essays would appeal less to the narrow interest of the specialist than to "those who have retained a true sense for the marvelous, and who are capable of discerning in the ordinary flow of events the point at which the mundane world gives way to quite another dimension" (1957b, 164).

With this change of orientation, Eiseley placed himself in a long and rich tradition of English and American natural history writers—including Gilbert White, Richard Jefferies, W. H. Hudson, Ralph Waldo Emerson, and Henry David Thoreau. Each of these writers, though prescientific, recorded his impressions of nature in a precise, distinctive, and personal style. Although obviously related to an older pastoral tradition, natural history writing differs by avoiding the sentimentality, the stock nature descriptions, and the "pathetic fallacy" typical of literary pastoralism. Instead, the natural history essay gains vividness and accuracy from the influence of empirical science. It is both personal and factual, balancing objectivity with delight. Yet the literature of natural history and modern scientific writing are also distinguishable in several important ways. The essence of science is to be quantitative and experimental in approach, whereas the art of natural history is to observe, appreciate, and record natural phenomena as they appear. This is not to deny the importance of observation in science, since a discovery often begins with the recognition of "broken symmetry," when a scientist notes the exception in the pattern. But science extends beyond observation to the making and testing of hypotheses.

In "The Enchanted Glass" (1957a), Eiseley eloquently defines the borders between literature and science.[2] The study of nature can be either factual or contemplative, depending upon the mindset of the person involved. Eiseley compares the two mindsets, finding the Baconian "severely experimental, unaesthetic and empirical," and its opposite "literary, personal and contemplative" (1957a, 478). Then he shows how they differ in attitude and purpose: the scientific attempting to discover the "underlying laws" that govern the natural world, and the contemplative seeking to record personal responses to that world. Description and analysis alone cannot provide a full understanding of the natural world, according to Eiseley, because "when the human mind exists in the light of reason, and no more than reason, we may say with certainty that man and all that made him will be in that instant gone" (1957a, 482). There is also a need for contemplative natural history, an approach that "contains overtones of thought which is not science, nor intended to be, and yet without which science itself would be poorer" (1957a, 482). Contemplative natural history offers what Eiseley calls a "natural history of the human soul" (1957a, 480). This human response to the world is especially important, he notes, in an age that does not lend itself to contemplation. A third approach is also possible, however. Besides straightforward science and contemplative natural history, there is a more personalized scientific literature that reflects a labor of love, in which fact and knowledge are balanced by affection. Eiseley valued both these approaches for their "synthesis of knowledge and emotional insight" (1957a, 492).

The natural history writer is primarily interested in recording his aesthetic response to the natural world. As an outdoorsman or field naturalist, he shows a general interest in the plants and animals of his native region.

Always he cultivates an affectionate sense of locale, an appreciation of a particular physical landscape, whether it be White's Selborne, Jefferies's Swindon, Hudson's pampas, Emerson's Concord, or Thoreau's Walden. The world of the natural history writer appears through the medium of a distinct personality, a recognizable "voice," while the modern scientist strives to maintain strict objectivity, refining the observer out of existence. This myth of complete scientific objectivity, epistemologically dubious, has encouraged the writing of scientific papers in dull, lifeless prose that is supposed to appear unbiased and detached. Yet one cannot efface point of view so long as there is a person doing the observing or recording the data.

This dry, passive scientific style is the antithesis of art—it is the voice of people without faces or personalities, an echo from the dissection room. Fortunately there are now alternatives to this "official" style of science, due in large degree to the influence of such articulate spokesmen for science as Eiseley, Jacob Bronowski, Robert Jastrow, and Lewis Thomas—who do not hesitate to express their ideas through a personal medium. Many of them have returned to the natural history essay or other varieties of the essay form as a way of expressing their response to their work—their delight and pleasure—in an informal but accurate manner.

Not that Eiseley was an apostate scientist; it was simply that he could not express what he felt most deeply within the framework of the impersonal scientific style. In *The Advancement of Learning* (1605, 88), Sir Francis Bacon classifies scientists into two general categories, the "miners" (or researchers) and the "smiths" (or refiners). Eiseley was clearly one of the latter, gifted with a synthesizing, metaphoric mind that constantly sought connections between fact and imagination. He possessed the poetic vision to perceive the natural history of life's emergence as an epic event—a spectacle of prehistory—heretofore described in flat, expository language, but capable of being recast as a literary narrative. But it took a special kind of talent to accomplish this, one more akin to the Victorian literary naturalist than to the modern specialist, someone not intimidated by the "two cultures" division or the fear of bridging disciplines. Eiseley, of course, had long been pursuing separate careers as a poet and a scientist, with the hope of eventually combining them in some literary form. As he commented to a colleague's inquiry about encouraging good writing among scientists, "I always had a joint interest in English literature and science so that the 'two culture problem' never concerned me and I was never conscious of it except to the degree that I have been castigated by nonliterary colleagues in science."[3]

Although C. P. Snow (1959) blamed the "two culture" crisis largely on literary intellectuals' unwillingness or inability to master basic scientific principles, the issue is really more complicated than that.[4] The split is of recent origin, caused in part by the explosion in scientific knowledge. The ideal of a common culture has given way to a range of specialized audiences—literary and scientific—with few overlapping interests. The contemporary scientist necessarily addresses his or her peers rather than a general public. The Victorian scientist, on the other hand, had to be a man of letters because he wrote for a general rather than a specialized audience—one he could not depend on to extract his argument from his charts and diagrams if his prose was murky or obscure. Darwin, Wallace, T. H. Huxley—the Victorian scientists and writers whom Eiseley so admired—mastered the gift of expressing complex ideas simply in order to reach their audiences, who were skeptical of evolutionary theory. Even Darwin's prose, so ponderous in places, often surprises the reader with passages of vivid imagery and persuasive skill. The study of natural history was respected by these men, who were not constrained by our present-day mania for specialization. As Eiseley notes in *The Night Country*:

> Even though they were not discoverers in the objective sense, one feels at times that the great nature essayists had more individual perception than their scientific contemporaries. Theirs was a different contribution. They opened the minds of men by the sheer power of their thought. The world of nature, once seen through the eye of genius, is never seen in quite the same manner afterward. A dimension has been added, something that lies beyond the careful analysis of professional biology. (1971, 142)

In many ways Eiseley resembles these gifted Victorians, naturalists with a literary bent who forever changed our understanding of our place in the natural world. An eloquent example of Victorian science is Thomas H. Huxley's *On a Piece of Chalk*, a lecture on geological time and evolution given in 1868 to the workingmen of Norwich. Eiseley's introduction to a recent edition (1967) of Huxley's lecture reveals much about his divided loyalty between literature and science. He speaks of "the two faces" of Huxley—the honest and tireless champion of science and the driven, defeated man "rendered intellectually impotent before space, time, and the unknowable" (Huxley [1868] 1967, 10). Like Huxley, Eiseley retained throughout his life "a lingering poetic eloquence, a fondness for the literary essay turned to scientific purposes." He praises Huxley and Agassiz, two men who disagreed about evolution, but who, because they agreed about "the necessity of writing lucidly upon scientific subjects for the layman shared a common tradition" (Huxley [1868] 1967, 10).

Even in his thematic concerns Eiseley shares much with these Victorian naturalists. As in their work, time and evolution have been the dominant themes in virtually everything Eiseley has written. But unlike Darwin, Eiseley did not write cautiously to introduce a disturbing and even heretical theory bearing on the origins of man. Instead he explored the poetic implications of the evolutionary metaphor through a series of visionary extensions in time, imaginatively re-creating those natural events that led to the emergence of man; but his terms are far more tempered and compassionate than Tennyson's "nature, red in tooth and claw." Still, there is something distinctively Victorian about Eiseley's intellectual temperament and his dominant interest in the history of life and its implications for what Bronowski optimistically called "the ascent of man." Eiseley was not so confident of the direction this evolutionary metaphor suggested, but if it was an "ascent," then it was a movement toward greater compassion and kinship with all forms of life rather than simply an advance in technological and manipulative skill.

THE CONCEALED ESSAY

Despite these affinities with Victorian men of science, it would be a mistake to view Eiseley's work as derivative or anachronistic. His visionary power, his animistic descriptions, and the range of his philosophical speculations distinguish him from his Victorian predecessors. Eiseley was less optimistic about the prospects of science than either Darwin or Huxley, so that a sense of diminishment pervades his work. A strong romantic tendency reinforces this pessimistic tone, particularly in his various self-dramatizations: as the neglected child, the impoverished student, the wandering bone hunter, the midnight scholar, the weary insomniac. While this attitude is largely temperamental, it may also reflect, especially in his most melancholic or somber moods, the influence of his diverse reading—Edgar Allan Poe, Robert Burton, and Sir Thomas Browne, for example. It may also be the mood in our time.

As an undergraduate at the University of Nebraska, Eiseley had studied the nineteenth-century essay under Professor Kenneth Forward, who may have been the source of Eiseley's later remark that "there are no perfect essays, only well-composed ones" (Eiseley Collection, undated entry). Throughout his life, his favorite authors remained Samuel Coleridge, William Hazlitt, Thomas De Quincey, Charles Lamb, Francis Bacon, and Sir Thomas Browne—all masters of the essay who loved obscure and arcane references; elegant practitioners of an ornate style who emphasized tone and mood in their writing. Eiseley may have learned from Bacon the habit of balancing scientific or scholarly ideas with vivid metaphors, although in this respect he is far more candid than Bacon, especially when he mentions the troubled and painful memories of his childhood. Retaining Bacon's philosophical sweep and scientific vision, Eiseley tempers it with a baroque, introspective style more reminiscent of Robert Burton, Thomas Browne, Abraham Cowley, and the later Thoreau. Certainly Eiseley shared Browne's interest in quaint and antiquarian lore, with its emphasis on the remote and mysterious, though Browne also served as an intriguing model of what he himself called a "great and true amphibium," a man of science who was able to retain his religious faith (Eiseley 1958b, 7).

Each of these writers cultivated the essay in its purest form, as a mode of personal expression. Edmund Fuller has observed that "the word 'essay' is derived from 'essai'—a noun from a verb—the *attempt* to express oneself, in brief and highly personal terms, on some single subject" (1964, 9). Montaigne developed this prose form in the sixteenth century as a medium for what Fuller calls "reflective speculation and self-examination" (9), and Francis Bacon subsequently introduced it into the English literary tradition with the brilliant and concise formulations in his *Essays*, first published in 1597. The essay found a ready audience through the eighteenth-century newspaper, beginning with the *Tatler* and the *Spectator*. Under the influence of Samuel Johnson, Oliver Goldsmith, and Lord Chesterfield, the essay became a reasoned, reflective form, though it remained for the romantic writers to expand the scope of the essay as a personal or familiar form.

What Eiseley accomplished is virtually to invent a new genre—an imaginative synthesis of literature and science—one that enlarged the power and range of the personal essay. Eiseley, who was remarkably learned and well-read, a lifelong antiquarian and lover of old books, found the personal or familiar essay the most congenial form through which to express his wide range of interests. The literary essay turned to scientific purposes offered the ideal mode for such self-discoveries as he wished to offer, modest and unassuming, tentative and speculative without appearing dogmatic (1975, 177–78). Eiseley's style, though distinctive, seems to echo the ornate, mannered form of the seventeenth-century prose writers.

Eiseley's mastery of the personal essay developed gradually. As he explains in *All the Strange Hours* (1975), he returned to the familiar essay in the late 1940s after "a scientifically oriented magazine which had requested an article from me upon man's evolution reneged in favor of a more distinguished visitor to America" (177). Eiseley then decided to rework the rejected essay in a more literary fashion—"into what I now term the *concealed essay,* in which personal anecdote was allowed gently to bring under observation thoughts of a more purely scientific nature" (177, emphasis added). The "concealed essay" starts with a vivid anecdote or reminiscence and gradually expands it in a scientific or contemplative direction. The subject matter of the essay, whatever it may be, is framed or "concealed" by the personal approach, which serves as a rhetorical device to engage the reader's attention. Thus the "concealed essay" became for Eiseley a highly elaborate form, with frequent literary references and allusions, numerous quotations, multiple themes, and an interwoven structure of contemplative concerns. This casual and informal, though sophisticated, technique brings narrative and personal experience, essentially fictional and autobiographical tools, to bear on what is otherwise simply expository material— scientific fact and hypothesis. Eiseley justifies his style as adding personal interest without distorting the accuracy of the scientific material:

> That the self and its minute adventures may be interesting every essayist from Montaigne to Emerson has intimated, but only if one is utterly, nakedly honest and does not pontificate. In a silence which nothing could impinge, I shifted away from the article as originally intended. A personal anecdote introduced it, personal matter lay scattered through it, personal philosophy concluded it, and yet I had done no harm to the scientific data. (1975, 178)

Thus the familiar essay points out the connections between things that seem quite disparate. Beginning with a specific setting or thematic description, it creates a dramatic occasion and then dramatizes the self in that occasion. It takes readers on an intellectual journey and returns them with new insights. Along the way it maintains the voice of informed conversation on a common theme, blending literature, science, religion, and philosophy—a whole range of gratuitous information that demonstrates the common ground among seemingly unrelated topics and allows readers to see for themselves how they relate.

This approach may have been first suggested by Eiseley's anthropology lectures at the University of Kansas, where he taught a variety of courses in new subject areas. One of the lecture techniques he devised was to enliven his presentations by drawing on his field experiences and using them to illustrate his more abstract material. The same balance of the personal and the scientific distinguishes Eiseley's mature style. Part of his accomplishment as a writer is that his essays seem so artless, when, as he acknowledges in his autobiography, they were actually so highly contrived:

> In all the questioning about what makes a writer, and especially perhaps the personal essayist, I have seen little reference to this fact; namely, that the brain has become a kind of unseen artist's loft. There are pictures that hang askew, pictures with the outlines barely chalked in, pictures torn, pictures the artist has striven unsuccessfully to erase, pictures that only emerge and glow in a certain light. They have all been teleported, stolen, as it were, out of time. They represent no longer the sequential flow of ordinary memory. They can be pulled about on easels, examined within the mind itself. The act is not one of total recall like that of the professional mnemonist. Rather it is the use of things extracted from their context in such a way that they have become the unique possession of a single life. The writer sees back to these transports alone, bare, perhaps few in number, but endowed with a symbolic life. He cannot obliterate them. He can only drag them about, magnify or reduce them as his artistic sense dictates, or juxtapose them in order to enhance a pattern. One thing he cannot do. He cannot destroy what will not be destroyed; he cannot determine in advance what will enter his mind. (1975, 156)

What Eiseley has given us here is a formula for his work, a glimpse into the mind of the writer at work, and a statement of the forces that compelled him to write. The principle operating here seems to be a reenactment through memory, a heightened and dramatized account of past moments of intensified experience that eventually led to new and startling insights.

Memory, landscape, and visual imagination combine to shape many of Eiseley's most powerful narrative passages. The first chapter of *The Immense Journey* gives the reader an immediate sense of a powerful eidetic imagination at work, a method of composition, as Eiseley has noted, through "the pictures that haunt my mind" (1975, 154). In "The Slit" this visual imagination freely moves from the present landscape, the bleak uplands of western Nebraska, to a vertical escarpment, which serves as a point of departure for a metaphoric journey backward in time, accentuated by Eiseley's chance discovery of a primitive mammal's skull, staring at him, embedded in the sandstone wall. That moment of discovery serves in turn as the impetus for a second metaphoric leap, an imaginative extension backward to the early Tertiary period, where "cat and man and weasel must leap into a single shape" in a common ancestor (1957b, 5). While Eiseley chips away at the fossil skull, a third metaphoric transition takes place as he becomes aware of the marvelous dexterity of his fingers—"the human hand that has been fin and scaly reptile foot and furry paw" (1957b, 6). Using the poetic device of the synecdoche, a substitution of the part for the whole, he achieves a remarkable compression of time, and through the leap of his imagination from fact to symbol he emphasizes the fortuitous chance that marks so much of evolutionary change.

Every image, every transition, is carefully selected to remind us of the enormous role of chance and contingency along the road of that vast "caravan" in time that has produced humankind (1957b, 12). Time remains, in Eiseley's words, "a dimension denied man" (1957b, 11), but one that he may yet learn to comprehend, through the powers of scientific imagination. Most important, though, the prehistoric past serves for Eiseley as a symbol of something remote and unreachable, a "middle border" of awe and wonder (1975, 25), a tonic to the jaded imagination. There is scarcely another writer for whom the remote, geological past is so imaginatively vivid as it is for him.

Eiseley recognized that the concept of geological time is one of the most difficult scientific notions for the ordinary person to comprehend. As he was to argue later in *The Firmament of Time* (1960), "we have difficulty in visualizing the age-old processes involved in the upheaval of mountain systems, the advance of continental glaciations or the creation of life" (3). Less than two hundred years have passed since the discovery of the vast age of the earth. Thus, in *The Immense Journey*, Eiseley is forced to devise a new mode of visionary natural history to convey his impression of the protean quality of life as viewed from a geological perspective. In his imaginative synthesis of literature and science, he combines narrative and hypothesis, fact and feeling, metaphor and exposition. The result is a new style of scientific literature and a new literary genre, an expansion of the personal essay for scientific purposes.

RECEPTION OF *THE IMMENSE JOURNEY*

In an important passage from "The Slit," Eiseley speaks of being restrained by the conventional methods of science because they do not allow adequate freedom to express his sense of the miraculous in the natural world. As he often implies in wry anecdotes about the disapproval of some of his colleagues, Eiseley recognizes the scientific method as a kind of contemporary orthodoxy, which he threatens, as Bacon did in his attack on medieval Scholasticism, by questioning its scope and limitations. As Eiseley warns his readers repeatedly, he is not a spokesman for conventional science. Instead, he is a poet-shaman, a wizard-alchemist who adopts the disguise of a changeling and calls his reverence and compassion for life "scientific" for the sake of his professional reputation. In another early essay, "Obituary of a Bone Hunter" (reprinted in *The Night Country*), he explains that in a series of three temptations, which our age would call "opportunities," he refused the possibility of fame and reputation either to protect another living creature or to maintain his personal integrity. Even in this early essay his composure over the loss of fame suggests the philosophical direction his work would take, the insight or idea becoming more important than the physical artifact. In each of the three incidents he gives up his quest to find evidence of early man in North America. The reasons why he turns away from these promising sites have little to do with the potential value of the finds. Each time there is some personal scruple that cannot be violated; a fear of spiders in close quarters, an unwillingness to disturb an owl's nest, a refusal to cooperate with the demands of an eccentric.

Given Eiseley's explicit literary intentions and his repeated disavowals of conventional science, it is not surprising that some of his colleagues in anthropology were skeptical about his work, but the lack of literary recognition for his accomplishments as an essayist and a prose stylist is more difficult to explain. Book reviewers have generally been enthusiastic, and his essays have been widely reprinted and anthologized, but Eiseley has still received relatively little serious critical attention. One of the few exceptions to this neglect has been Van Wyck Brooks's tribute, in the last volume of his autobiography, *In the Shadow of the Mountain: My Post-Meridian Years*:

> I have long believed that the best writers are now the writers of natural history who are ignored commonly in critical circles because they are concerned with permanent things outside of the changing human

world that interests the novelists and most of the poets. . . . Otherwise there would be no writers more critically esteemed than Henry Beston or Rachel Carson or Loren Eiseley, who has related, in *The Immense Journey*, the ascent of man from his dark stairwell. Why are these writers of natural history now called popularizers of science as if their style went for nothing, as if this were the bottom rung of the ladder of science instead of an upper rung of the ladder of art? (1961, 21)

Brooks is undoubtedly correct in his assessment of the lack of critical interest in natural history writers. Yet *The Immense Journey* and Eiseley's subsequent books have enjoyed so wide a readership that he is assured a prominent place among such modern natural history writers as Aldo Leopold, Joseph Wood Krutch, Marston Bates, and Annie Dillard, all accomplished stylists who blend their natural descriptions with acute philosophical observations.

NOTES

1. One of the most intemperate criticisms of Eiseley's style occurs in John Buettner-Janusch's review of *The Firmament of Time*. He faults Eiseley for his "sentimental and trite" writing, his "fevered prose, overblown metaphor, and sentimental twaddle" (1963, 9).

2. Two of Eiseley's best discussions of the difference between natural history and science can be found in "The Enchanted Glass" (1957a, 478–92) and *The Night Country* (1971, 127–48). For further reading, see Joseph Wood Krutch's introduction to the genre of the natural history essay in his lengthy prologue to *Great American Nature Writing* (1950).

3. These remarks may be found in a 2 May 1966 letter from Eiseley to Edwin G. Boring of Harvard (in the Eiseley Collection, University of Pennsylvania Archives). Eiseley was director for a time of the Richard Prentice Ettinger Program for Creative Writing (in the sciences) at the Rockefeller Institute in New York.

4. C. P. Snow's *The Two Cultures and the Scientific Revolution* (1959) contains one of the earliest and best discussions of the cultural and intellectual division between the humanities and the sciences. Eiseley (1964) dismisses this gap, however, as more imagined than real.

REFERENCES

Amiel, H. F. [1883–84] 1935. *The Private Journal of Henri Frederic Amiel*. Trans. V. W. Brooks and C. V. W. Brooks. New York: Macmillan.

Bacon, F. [1605] 1974. *The Advancement of Learning and New Atlantis*. Ed. A. Johnson. Oxford: Clarendon Press.

Bergson, H. 1944. *Creative Evolution*. Trans. A. Mitchell. New York: Random House.

Brooks, V. W. 1961. *In the Shadow of the Mountain: My Post-Meridian Years*. New York: E. P. Dutton.

Buettner-Janusch, J. 1963. The Firmament of Time. *American Anthropologist* 65: 9.

Eiseley, L. 1957a. The Enchanted Glass. *American Scholar* 26: 478–82.

———. 1957b. *The Immense Journey*. New York: Random House.

———. 1958a. An Evolutionist Looks at Modern Man. *Saturday Evening Post*, 26 April.

———. 1958b. *Darwin's Century*. Garden City, N.Y.: Doubleday.

———. 1960. *The Firmament of Time*. New York: Atheneum.

———. 1964. The Illusion of the Two Cultures. *American Scholar* 33: 387–99.

———. 1971. *The Night Country*. New York: Charles Scribner's Sons.

———. 1975. *All the Strange Hours*. New York: Charles Scribner's Sons.

———. Eiseley Collection. University of Pennsylvania Archives.

Fuller, E., ed. 1964. *The Great English and American Essays*. New York: Avon Books.

Huxley, T. H. [1868] 1967. *On a Piece of Chalk*. Intro. L. Eiseley. New York: Charles Scribner's Sons.

Krutch, J. W. 1950. Prologue. In *Great Natural History Writing*. New York: William Sloane.

Snow, C. P. 1959. *The Two Cultures and the Scientific Revolution*. New York: Cambridge Univ. Press.

In Search of the Exact Location of the Soul:

Richard Selzer and the Rhetoric of Surgery

CHARLES M. ANDERSON

In 1975, Richard Selzer described the "two special factors in the development of a doctor" as guilt and self-protection:

> We doctors swim in a sea of guilt. . . . A surgeon is created on the wreckage of a legion of patients who have survived his mistakes. Whereas we hate our guilt, and it is painful for us to endure day after day, we need it. . . . The other related factor is self-protection. By God, we have to protect ourselves! . . . We must surround ourselves with callouses so that we cannot feel the sharp prick and sting of what we do. (Trautmann 1981, 14)

Nonetheless and simultaneously, the surgeon must also be aware that the inert "object" upon which he "works" is a human being and that he is inflicting a great deal of pain in his attempts to cure that human being.

This paradox is articulated in a conversation recorded in Joanne Trautmann's *Healing Arts in a Dialogue*, in which one participant in the series of retreats around which the book is organized observes that "it is dangerous to think and ponder and feel every time you cut a patient, to see when you lose a patient that you've lost a life. To have this happen again and again is the rock getting worn away. It's a human life getting worn away." To which Selzer replies, "I know that to be a fact. And yet woe betide the patient whose surgeon forgets for one moment that there is a living human being under those sheets with the same conflicts, wants, needs, and dreams as he himself has. Then the act of surgery takes on a kind of violence to me, a sense of cruelty, and a bloodiness that it ought not to have. Then one runs the risk of embruing one's hands with gore" (Trautmann 1981, 106).

Recognizing the humanity of the patient and identifying one's "conflicts, wants, needs, and dreams" with the inert form on the table, a form that one must shortly lay open with a knife, is no easy task. It demands that the surgeon lay himself or herself open to the whole of the medical event, experiencing it from the technical point of view, from the illness perspective of the patient, and from the surgeon's own human perspective. In addition, the surgeon must act in this context of conflicting, if not contradictory, imperatives. He or she must assume control, knowing all the while that the unpredictability of disease fundamentally undermines any attempt at control over the events of the surgical theater. Such a collision of imperatives tears at the surgeon, forcing him or her into a rhetorical situation of astounding complexity. As Selzer puts it, "It is the nature of creatures to live within a tight cuirass that is both their constriction and their protection. . . . The surgeon cannot weep. When he cuts the flesh, his own must not bleed. Here it is all work. Like an asthmatic hungering for air, longing to take just one deep breath, the surgeon struggles not to feel. It is suffocating to press the feeling out" (1976, 101). The quality of the pressure on the surgeon is perfectly expressed by the image of the asthmatic, whose need to breathe is joined by metaphor to the surgeon's equally desperate need not to feel. The result is a picture of someone struggling to inhale (the asthmatic) and not to inhale (the surgeon) at the same time. The physical sensation is one of impossible stasis, involuntary muscles straining against voluntary ones for control. This surgeon is not swimming in a sea of guilt, he is drowning in one.

For Selzer, the rhetorical options that protect his colleagues—the armor of indifference, the bifurcation of professional and private lives, the detachment of concern—proved insufficient to absorb, deflect, or resolve the pressures of his profession. In an interview published in 1984, Selzer explains his particular solution:

> I knew I was going to be a doctor from early childhood. My father was a general practitioner in upstate New York who died when I was 12. . . . When I reached the age of 40, 39 or 40, quite abruptly I felt a certain need to do something else, and I wasn't quite sure what that was going to be. I felt restless, a kind of agitation, if you will, of the spirit and almost a malaise. And rather without prior intention I took pencil and paper and wrote. I wrote a short story that was a retelling of the Jonah and the Whale myth, and suffice it to say that when I read it over three hours or four hours later I knew at once that I was

a writer. I was in fact healed, the physician had healed himself, I was calm. (Schuster 1984, 85)

In the interview Selzer does not make much of his malaise, nor does he expand upon the nature of the healing that writing "Jonah and the Whale" produced. But in unpublished notes from which he composed many of his early speeches, he presents both in a clearer, more urgent way:

> [I write] to domesticate my terrors. . . . I do it to ward off disease, fend off death, to give pain a name. I think I should have died at the age of forty if I had not begun to write. For me to write is to transform all of my helplessness and despair as a surgeon into an affirmative act of creation. . . . [A] work of Art stands between the artist and extinction. Art . . . provides a way of reshaping pain and the fear of death into a harmonious, healthy, living thing.

In another place in these same notes, Selzer says, "My writing is my life vouchsafed."

While Selzer's early writing provided the healing necessary for him to continue practicing medicine, its ultimate value and importance are not limited to his personal situation. As he began to view his own experience through the various lenses of fiction and the essay, he began to see that indifference, bifurcation, and detachment need not be the only stances a physician might take. Selzer discovered important and meaningful alternatives to scientific medicine, alternatives that reach back to the very beginnings of medicine itself and draw on powerful processes of symbolic, linguistic, and poetic transformation. The result is not a critique of contemporary medical practice but a revision of contemporary medical discourse. While there are any number of places in Selzer's writing that demonstrate his alternative discourse, no place displays the means by which it came into being better than the opening section of his first book of essays, *Mortal Lessons: Notes on the Art of Surgery* (1976). In the first essay, "The Exact Location of the Soul," Selzer develops many of the terms and concepts that dominate his discourse and inform the rhetorical vision his writing brings to the act of healing, a vision that moves from stasis, bewilderment, and pain to exuberance, understanding, and transcendence.

From the opening sentence of "The Exact Location of the Soul," there is poetic tension:

> Someone asked me why a surgeon would write. Why, when the shelves
> . . . sag under the deadweight of books. To add a single adverb is to

risk exceeding the strength of the boards. A surgeon should abstain. A surgeon, whose fingers are more at home in the steamy gullies of the body than they are tapping the dry keys of a typewriter. A surgeon, who feels the slow slide of intestines against the back of his hand and is no more alarmed than were a family of snakes taking their comfort from such an indolent rubbing. A surgeon, who palms the human heart as though it were some captured bird. (1976, 15)

Shelves threaten to break, steamy gullies contrast with dry keys, intestines—or is it hands?—become snakes, and the heart, a captured bird. Coming as they do at the beginning of the essay, which is also the beginning of Selzer's first collection of essays, there is no context to provide clues to help the reader unravel the meaning of Selzer's metaphors. And he does not explain them or answer the questions he poses. Instead, he uses their impact to challenge and to frustrate and to begin slowly altering his readers' expectations, inviting a radical reconsideration of their understanding of the surgeon, of the human body, and of the surgeon's relationship with that body. The language gives pause. Weighted adverbs? Snakes? Birds? It startles and unsettles through what Paul Ricoeur in *The Rule of Metaphor* calls "a bringing together of terms that first surprises, then bewilders, and finally uncovers a relationship hidden beneath the paradox" (1984, 27).

The second paragraph continues to question the surgeon's motives. "Is it vanity that urges him? . . . Is it for money?" The language of Selzer's first paragraph suggests that this is not the case, and he is quick to affirm that he is after something entirely different:

No. It is to search for some meaning in the ritual of surgery, which is at once murderous, painful, healing, and full of love. It is a devilish hard thing to transmit—to find, even. Perhaps if one were to cut out a heart, a lobe of the liver, a single convolution of the brain, and paste it to a page, it would speak with more eloquence than all the words of Balzac. Such a piece would need no literary style, no mass of erudition or history, but in its very shape and feel would tell all the frailty and strength, the despair and nobility of man. (1976, 15)

Most people believe surgery to be a way to recover from a disease, to involve pain, for sure, but certainly not to be party to either murder or love. In what way could surgery, with all its technical and scientific apparatus, be likened to a ritual? The concrete images Selzer uses in his initial figuration struggle to establish their meaning in a broader context in which such figuration is not only unexpected but almost certain to be condemned. One reviewer, for example, observed that he felt unsettled by descriptions of modern medicine couched in language more appropriate to a period of medical history in which barbers, not doctors, performed primitive, brutal surgical procedures (Morgan 1979, 53). He is right, and that is precisely Selzer's point: to disturb and to unsettle his readers by increasing the tension among their expectations, the literal view of scientific medicine, and his metaphoric representation of surgery so that he may move beyond the "facts" of surgery toward its meaning.

The problem for the surgeon who would attempt this feat is that it may be an impossible task, that the only way to tell the "frailty and strength, the despair and nobility of man," may be to publish parts of the body itself, cut out and pasted to the page. Selzer immediately recognizes that such a "synecdochical" publication is "preposterous" but fears "that is what it may require to reveal the truth that lies hidden in the body." Others have tried to write it out—Rabelais, Chekhov, William Carlos Williams—but none of their peculiarly violent "undressings" or heroic "assaults have wrested it free" (1976, 15–16).

In the third paragraph Selzer explains why his predecessors have failed and sets forth his own "program." "I have come to believe," he writes,

> that it is the flesh alone that counts. The rest is that with which we distract ourselves when we are not hungry or cold, in pain or ecstasy. In the recesses of the body I search for the philosopher's stone. I know it is there, hidden in the deepest, dampest cul-de-sac. It awaits discovery. To find it would be like the harnessing of fire. It would illuminate the world. Such a quest is not without pain. Who can gaze on so much misery and feel no hurt? (1976, 16)

In the opposition between pain and ecstasy, pain serves to ground or to root one end of human experience in the physical, a grounding Selzer finds central to the mortal condition, but ecstasy serves to pull away from the merely physical by connecting or relating the physical to the realm of feeling, which Ricoeur suggests is a cognitive, metaphoric, or poetic response to the physicality of bodily sensation (Ricoeur 1978, 155–59). Even in its medical usage, *ecstasy* signifies a condition of both the flesh and the mind. In the *Oxford English Dictionary* (*OED*), the modern scientific or medical meaning is "certain morbid states of the nervous system, in which the attention is occupied exclusively by one idea." In other senses, *ecstasy* refers to "the state of being 'beside oneself,' thrown into a frenzy or stupor with anxiety, astonishment, fear, or passion," mystical trances in which the body becomes "incapable of sensation while the soul [is] engaged in the contemplation of divine things," and "an exalted state of feeling which engrosses the mind to the exclusion of thought" (*OED* 1973).

I focus so tightly on this word because it represents an important aspect of Selzer's technique as a writer. In our conversations in April 1984, Selzer talked about one of the means he employed to become a writer.

> I decided to study the English language as though I didn't know it, as though it were a foreign language. . . . I spent two years . . . learning the language, reading the dictionary . . . finding out the latitude of the language, which is enormous. The English language is unbelievable. The width of it, the depth of it; that you can use one word and have it mean ten different things. . . . It was a freedom, freedom I found.

This education in the language made Selzer particularly conscious of the power residing in the complex, interconnected, often contradictory meanings of individual words. In this particular case, *ecstasy* brings into the reader's consideration the trichotomy of mind, body, and soul and suggests a number of ways one might approach it. These include the elevation of the mind and its single idea over both the body and the soul; the mystical elevation of the contemplating soul, which moves one out of one's rational self and away from one's unfeeling body; and, finally, the elevation of feeling over the rationality of the mind and the spirituality of the soul. What Selzer accomplishes through the choice of this particularly "multivalent" term is the opening of a rich field of potential meanings that contains all of the elements with which he wishes to deal but does not limit him to any one of them. Instead, they act and interact in ways that further unsettle readers who search for clues and who become more and more engaged by the invitational ambiguities that continue to build around Selzer's redescription of surgery.

Selzer's announcement that he searches for "the philosopher's stone" works in a similar way. The stone is said to be a magical catalyst, which would, in the alchemist's hands, transmute lead into gold; in the philosopher's hands, redeem humanity and the universe (*American Heritage Dictionary* 1971); and in the hands of the surgeon, according to Selzer, it would "illuminate the world." The philosopher's stone has, in addition to its alchemical and other properties, "according to some, the power of prolonging life indefinitely, and of curing all wounds and disease" (*OED* 1973). However, it is not the stone's ability to cure *physical* "wounds and diseases" that Selzer seeks. It is the stone's ability to cure illness, the human perception and experience of "wounds and diseases," that he wishes to employ. For Selzer, the philosopher's stone, as a part of myth, is a language construct, not a tangible object. It serves to bring into the "field" of the essay considerations and concepts that cross the boundaries of myth, science, magic, medicine, language, and philosophy to create a free and protean complex of possible references. The world it promises to illuminate is the world of human suffering, a world constructed of language uttered in response to pain and ecstasy. These assertions are supported by two specific and critical points in the text of "The Exact Location of the Soul."

The first point is the question that immediately follows Selzer's discussion of the philosopher's stone: "Who can gaze on so much misery and feel no hurt?" (1976, 16). This deceptively simple question opens the essay to a host of rhetorical considerations by focusing specifically on the symbolic realm of illness rather than the physical aspect of disease and by pointing directly to the failure of the hierarchical, top-down rhetoric of "normal" medicine to adequately address the nonhierarchical, horizontal human relationships inherent in and necessary to the act of healing. In "normal" medicine, the physician is the one who is conditioned by his education and even by the language within which he does his healing not to feel, not to become a part of the procedures being performed. As Selzer expresses it for his own specialty, "surgery is masculine, tough, macho . . . feeling . . . a surgeon . . . mustn't feel. He must have and do, not feel" (1984 interview). But such a distancing, attractive as it might seem, is not acceptable, nor is it ultimately possible or even desirable, lest the surgeon "run riot on the populace" or "risk . . . embruing [his] hands with gore" (Trautmann 1981, 16). Selzer's question not only calls this conditioning to account, but in its very syntax, it invites the kind of participation that a rhetoric of identification in which patient and physician hold equally powerful and equally important positions makes possible.

The second point follows immediately after the question and serves to contextualize the essay so that readers can begin to locate "clues" to help them understand Selzer's meaning, clues that begin to resolve some of the tensions that have built through the essay. "Emerson has written that the poet is the only true doctor. I believe him, for the poet, lacking the impediment of speech with which the rest of us are afflicted, gazes, records, diagnoses, and prophesies" (1976, 16). The reference is to Emerson's essay "The Poet," which provides a number of essential concepts that clarify the exact nature of the illumination Selzer seeks.

In the first part of "The Poet," Emerson describes the philosophers of his time who had split the human soul or spirit from the physical body, who seemed "to have lost the perception of the instant dependence of form upon soul" (1960, 222). But, Emerson counters,

> we are not pans and barrows, nor even porters of the fire and torch-bearers, but children of the fire, made of it, and only the same divinity transmuted and at two or three removes, when we know least about it. And this hidden truth, that the fountains whence all this river of Time and its creatures floweth are intrinsically ideal and beautiful, draws us to the consideration of the . . . Poet, or the man of Beauty; to the means and materials he uses. (223)

In his speech notes Selzer describes his perception of this essential unity of spirit and flesh and of its place in his writing: "In my patients I find most accessible the marvelous proximity of spirit and flesh that is the only subject for the artist. It is impossible to separate a person's spirit from his body. . . . I have come to understand that the flesh is but the spirit thickened. Yes, this skinful of nerves and marrow is the jellied soul."

These passages suggest that there is no "truth that lies hidden in the body" that may be "heroically assaulted" and "wrested free," but that the illumination Selzer and his more manipulative fellow writers seek is to be found in the very inextricability of the "divine fire" and the "flesh." Soul and body, ecstasy and cold and hunger and pain, all elements of the mortal condition, are one and the same and are "intrinsically ideal and beautiful." Their discovery and representation compose the task of "the Poet, or the man of Beauty," whose materials are language and perception and whose means are metaphor and correspondence. For such a poet, pain, misery, and hurt, being elements of the mortal condition, may become the path-ways down which he can discover and experience, and the means by which he may transmit his discoveries and experiences to others in "records, diagnoses, and prophecies." Emerson calls this aspect of the poet's work "publication" and says,

> I know not how it is that we need an interpreter, but the great majority of men seem to be . . . mutes, who cannot report the conversation they have had with nature. . . . Too feeble fall the impressions of nature . . . the rays or appulses have sufficient force to arrive at the senses, but not enough to reach the quick and compel the reproduction of them-selves in speech. The poet is the person in whom these powers are

in balance, the man without impediment, who sees and handles that which others dream of, traverses the whole scale of experience, and is representative of man, in virtue of being the largest power to receive and to impart. (1960, 223)

In Selzer's operating theater, the "rays or appulses" of nature are intentionally weakened, thwarted by anesthetics of one kind or another, turning both the surgeon and the patient into "mutes." "The patient," writes Selzer in his speech notes, "must be anesthetized in order that he feel no pain. The surgeon too must be anesthetized in order that he be removed from the emotional heat of the event, in order that grief or passion not weaken his resolve." But, he continues, while the surgeon practices his science poetically, there is a third party present who "is awake. . . . He sees and hears everything, censors nothing. It is his job to report the event back to the world." This third party is Emerson's Poet, whose "sign and credentials . . . are that he announces that which no man foretold. He is the true and only doctor; he knows and tells; he is the only teller of the news for he was present and privy to the appearance which he describes" (Emerson 1960, 225). This doctor does not offer the patient a cure for the diseases of the body, but he does offer "something equally important—a moment of revelation" (Selzer 1983), a moment of symbolic illumination, which is the only cure for illness.

The surgeon who would become such a doctor "stands among partial men for the complete man, and apprises us not of his wealth, but of the common wealth" (Emerson 1960, 223). His particular aim and purpose are to speak from his vantage point as one who is intimately familiar with the workings of the human spirit under the extreme conditions created by disease and death and recovery to show us our strengths and weaknesses, our despair, and our nobility. The surgeon-poet thus would reach far beyond the boundaries of the doctor-patient relationship and draw the rest of humanity into the world of his work.

This is precisely what Selzer proceeds to do in the second section of "The Exact Location of the Soul" by presenting one of *Mortal Lessons'* "representative anecdotes," a story that, in Kenneth Burke's terms, provides Selzer with a suitable vocabulary through which to communicate his perceptions of the reality with which his writing is to deal (1974, 59). In Ricoeur's terms, it is a metaphorical story that "addresses itself to deeply rooted potentialities of reality to the extent that they are absent from the actualities with which we deal in everyday life under the mode of empirical control and manipulation" (1978, 154–55). In addition, the anecdote serves to create in the reader's mind a tension that far surpasses the perplexity of Selzer's earlier questions and problems, a tension that becomes an apparently unresolvable horror.

"I invited a young diabetic woman to the operating room to amputate her leg" (1976, 16). The active voice of the story's opening sentence, in contrast to the passive form of the normal scientific sentence, puts the surgeon explicitly into the position of agent, making him not only the person performing the surgical procedure but also putting him into a position to accept responsibility for it. The verb "invited" serves to extend the habitability of Selzer's syntax by bringing his patient into it via a rhetoric that does not force or manipulate her. To invite is to recognize and, at least implicitly, to accept the other's right to decline the invitation. At the same time, "invited" builds tension within the sentence because it is not an appropriate verb through which to express an action connected with an operating room, and certainly not one appropriate to an amputation. Aware of the dissonance, the reader seeks textual clues that might help reduce the tension and is thereby "invited" into the operating theater as well.

The story that follows is one of intense, unrelenting pain, of a "great shaggy black ulcer" that ate away at the flesh of the diabetic woman, who could not see it, "for she was blind as well." For over a year, the surgeon worked to save her by trimming away "the putrid flesh." "There is no pain like that of the bloodless limb turned rotten and festering. There is neither unguent nor anodyne to kill such a pain yet leave intact the body." So the surgeon and his patient must finally surrender to the necessity of an amputation. "At last we gave up, she and I. We could no longer run ahead of the gangrene. We had not the legs for it. There must be an amputation in order that she might live—and I as well. It was to heal us both that I must take up knife and saw, and cut the leg off. And when I could feel it drop from her body to the table, see the blessed space appear between her and that leg, I too would be well" (1976, 16–17). This passage is rhetorically important because of the movement from the "I" of the surgeon acting and the "she" of the patient being acted upon to the "we," the plural first-person pronoun, which brings them together in a single word, creating a single, consubstantial verbal entity. In Burke's rhetoric, this consubstantiality is essential to identification and the cooperative action that may arise from it (1969, 19–46). Here, it allows them both to "run ahead of the gangrene" and then to give up, first as consubstantial "we" and then as individuals, "she and I." The subsequent shift back into scientific syntax—"There must be an amputation in order that she might live"—serves to recognize and defer to the presence of necessity, an undeniable presence at the very center of this medical event, a presence expressed in a scientific sentence that empowers the surgeon to act and helps him and his patient to distance themselves from the events in which they are compelled to play a part. The surgeon, however, is not distanced from the patient. He has taken upon himself something of his patient's illness, for he adds, "and I as well. It was to heal us both." In his year with this patient he has "become" this patient through the relational power of the rhetoric of identification. At the same time, he is not the patient, because it is he who "must take up knife and saw, and cut the leg off" to create a "blessed space." This paragraph is one of Selzer's very best representations of the collision of hierarchical and horizontal rhetorics in surgery; but in this text, the collision involves more than a surgeon and his patient. There is another participant in these painful verbal events.

To this point, the reader has read the surgeon's report of what has already happened, a disturbing enough set of events, but distanced in time and place by past and conditional verb tenses. In the third paragraph of the anecdote, suddenly, "Now it is the day of the operation. I stand by while the anesthetist administers the drugs, watch as the tense familiar body relaxes into narcosis" (1976, 17). The "blessed space" that had separated the reader from the events being recounted is gone, and the reader, too, is in the operating room, transformed into a third party similar to the one Selzer described earlier, the person who is "awake," who "sees and hears everything, censors nothing." "I turn then to uncover the leg. There, upon her kneecap, she has drawn, blindly, upside down for me to see, a face; just a circle with two ears, two eyes, a nose, and a smiling upturned mouth. Under it she has printed SMILE, DOCTOR. Minutes later I listen to the sound of the saw, until a little crack at the end tells me it is done" (1976, 17–18). Selzer's final "I" is both the surgeon and the witnessing reader listening "to the sound of the saw" and to the "little crack at the end." It is a consubstantial "I" that allows no space between surgeon and reader, who are distanced from the amputation only by the "blessed space" of a sentence whose syntax allows them to "listen to the sound of the saw" rather than to be the one who saws.

What began as a surgeon's anecdote has become a bewildering, hurt-filled experience for the reader, who feels, in Ricoeur's verbal sense of the term, what both surgeon and patient feel and who is thereby inserted into the reality depicted by Selzer's language, a reality in which the misery of a festering limb and an amputation seems too great to be countered by "just a circle with two ears, two eyes, a nose, and a smiling upturned mouth." Yet, out of such pain grows Selzer's alternative to medical discourse, an alternative that, even in the despair of a rotting leg that must be amputated, sees and reports "Beauty" and "love."

> So, I have learned that man is not ugly, but that he is Beauty itself. There is no other his equal. Are we not all dying, none faster or more slowly than any other? I have become receptive to the possibilities of love (for it is love, this thing that happens in the operating room), and each day I wait, trembling in the busy air. Perhaps today it will come. Perhaps today I will find it, take part in it, this love that blooms in the stoniest desert. (1976, 18)

Perhaps. But how is one to reconcile the horror of "a little crack at the end" with "Beauty" and "love"? Emerson again provides clues in "The Poet":

> Wherever snow falls or water flows or birds fly, wherever day and night meet in twilight, wherever the blue heaven is hung by clouds or sown with stars, wherever are forms with transparent boundaries, wherever are outlets into celestial space, wherever is danger, and awe, and love—there is Beauty, plentious as rain, . . . and though thou shouldst walk the world over, thou shalt not be able to find a condition inopportune or ignoble. (241)

That is, no matter what conditions the poet faces, he will, if he is open to it, find beauty therein. "He is the poet [who] shall draw us with love and terror, who sees through the flowing vest[ment] the firm nature, and can declare it" (Emerson 1960, 238).

Though Emerson provides clues to understanding Selzer's sense of beauty, Emerson's abstract, romantic definitions undergo important transformations when "nature" becomes the grotesque and shocking universe of sickness, deformity, and death within which the surgeon operates and about which Selzer writes. Emerson might have accepted these transformations, recognizing as he does in "The Poet" that "thought makes everything fit for use," that "what would be base, or even obscene, to the obscene, becomes illustrious, spoken in a new connection of thought" (229), and in "Nature" that "even the corpse has its own beauty" (27). But Emerson could not have anticipated, imagined, or foretold the depth or degree of the transformation because, while his nature always contains disagreeable or foul objects, his images and metaphors and the perceptions they convey are overwhelmingly agreeable and fair.

The disagreeable, however, is not merely part of Selzer's nature; it inevitably, necessarily, occupies center stage: "The grotesque I accept as a matter of course. It is part deformity, part humor and part hallucination. Nature makes no mistakes. Everything, beautiful or ugly, has its cause and purpose. There is not one natural object in existence that ought not to be" (Selzer, undated Dallas address, 4–5). In his speech notes Selzer further refines and articulates his understanding of beauty and ugliness in his work.

> What is called ugliness in nature can become beauty in art. An ulcer, an amputated leg, a dwarf, dead flowers, blood spreading on a pillow. Baudelaire wrote of a festering corpse, eaten by worms, and imagined his own beloved mistress in a state of phosphorescent decay. Nothing can surpass his juxtaposition of beauty and disintegration. . . . To dilute your representation of pain, old age, or perversion in an attempt to protect your reader is to create ugliness.

But Selzer must not be seen as a mere sensationalist using the resources of his surgical profession to gather material to shock his readers' sensibilities. His aim is considerably higher.

In "Beauty and Pain: Notes on the Art of Richard Selzer," David Morris provides valuable insight into the specifics of Selzer's sense of beauty and its function. Morris begins with the assertion that the Western idea of beauty is derived largely from Edmund Burke's *Philosophical Enquiry into the Origin of Our Ideas of the Sublime and Beautiful* (1757):

> In the *Enquiry* beauty loses all its ancient (classical and Christian) associations with knowledge, truth, goodness, and wisdom. . . . "We must conclude," writes Burke, "that beauty is, for the greater part, some quality in bodies, acting mechanically upon the human mind by the intervention of the senses." Beauty, it turns out, is whatever produces the sensation of love, and what produces the sensation of love is anything appropriately small, graceful, varied, soft, colorful, undulating, delicate, mild, sweet, muted, or elegant. . . . Burke believed that he had advanced knowledge by discovering the psychological laws and social purpose of beauty—as the force which unites solitary individuals into a community of love—but his real achievement was to reduce the beautiful to a list of arbitrary forms and qualities. It occupies the exhausted category which we find reserved today for movie

stars, sunsets, rainbows, swimsuit competitions, old works of art, and little fuzzy animals who don't bite. (Morris 1980, 124)

While the possibility that it is beauty that "unites solitary individuals in a community of love" should not be rejected out of hand, Burke's domesticated, eviscerated beauty does not carry enough weight for Selzer, is not strong enough to embrace an amputation. Selzer's beauty is a more powerful, more ancient one capable of absorbing and transforming the painful spectacles the surgeon witnesses and in which he participates into the "possibilities of love." In fact, pain itself "is the unlikely medium through which Selzer attempts to recover some of the ancient resources of beauty":

> Selzer's is a concept of beauty that takes his reader back to Aristotle, who for over one thousand years provided the standard physiological account of pain, [and] offered in his *Poetics* a theory of tragedy in which pain and beauty achieve a subtle, necessary concord. Through a catharsis of pity and fear, pain ultimately carries the spectator to a state of intellectual, emotional, and aesthetic clarification. Knowledge and beauty replace the confused accidents of history, and it is pain which provides the essential raw material for the transmutations of tragic art. (Morris 1980, 126)

The feelings of pity and fear work by inserting the reader or viewer into the writer's representation or model of reality, thus creating a state of tension out of which "intellectual, emotional, and aesthetic clarification" arises. This process absolutely requires the reader or viewer to identify herself with the characters in the tragic situation, not in the sense of becoming them, but in the sense of becoming consubstantial with them in the way Kenneth Burke's "A" and "B" are both consubstantial and individual identities (1969, 20–21). The beauty and pain such an identification leads the reader to experience are imaginative, linguistic constructs accompanied by physical sensations, not purely physical sensations or bodily emotions. In one sense, these feelings are not "real" at all—viewers are not that which they view. In another sense, precisely because they are not what they view, the events are more real because they allow viewers to suspend ordinary, everyday references and relationships and thereby reveal, through the metaphorical process, "the deep structures of reality to which we are related as mortals who are born into this world and who dwell in it for a while" (Ricoeur 1978, 153).

The catharsis that accompanies such a revelation does not purge one's feelings as one might purge one's body of unwanted or dangerous substances; it purges in a very rare sense of the term that means "to reach out, extend, put forth," or "to issue forth" (*OED* 1973), making it possible for one to imaginatively leave one's everyday self and become a willing participant in the events of the representation. This is finally what Selzer's story of the blind diabetic woman's amputation invites readers to experience, a catharsis that brings them face-to-face with the horrors of the woman's mortal condition and with the "deep structure" implications of her transcendence of those facts—her crude drawing and the beautiful, silly, loving words "SMILE, DOCTOR."

The figure and its words are key elements that reveal that she has, in the midst of horror, reached past the surface structures of her suffering, past her own misery and into the deepest resources of human generosity to take upon herself a part of her surgeon's hurt. They are her way of leaving her self and touching his. Although they do not change the fact that she has lost her leg or the fact that she has suffered a year of pain to reach the amputation, they show that she has transmuted those facts into a human love that "blooms in the stoniest desert" of "a little crack at the end," a human love that finally becomes essential to Selzer's particular definition of the horizontal rhetoric of identification through which his surgeons may come to understand and to live with and within their daily practices.

In the ninth paragraph of "The Exact Location of the Soul," Selzer turns his attention again to the physician who would write. He says the doctor has been portrayed as a "figure of fun," and "is ripe for caricature." However, he believes "the truly great writing about doctors has not yet been done. I think it must be done by a doctor, one who is through with the love affair with his technique, who recognizes that he has played Narcissus raining kisses on a mirror, and who now, out of the impacted masses of his guilt, has expanded into self-doubt, and finally into the high state of wonderment" (1976, 18). In the *Oxford English Dictionary* (1973), the word *wonder*, upon which *wonderment* is based, carries oppositely charged meanings. The negative meaning is based on the wonder that has to do with "evil or shameful action . . . evil or horrible deeds . . . destruction, disaster . . . great distress or grief." This is what Selzer calls "self-doubt," what Emerson calls the perception of "daemons" (1960, 239). The positive meaning is "the emotion excited by the perception of something novel and unexpected, or inexplicable; astonishment mingled with perplexity or bewildered curiosity." What makes this a positively charged meaning is that it is precisely the same as the sensation that, according to Ricoeur's theory of the metaphorical process, precedes the recognition of a new metaphorical representation or a new model, perhaps the same sensation that generates such a representation, and very similar to the feeling that serves to decrease the distance between the viewer and what is viewed and to "insert" that viewer into the reality described by the new representation (Ricoeur 1978).

In Selzer's work, these two meanings of *wonder* interact to push the doctor to understand, resolve, and report the things he has seen. But the interaction is not simple; nor is it linear, from perception to action. It is, like the surgeon's response to his patients' illnesses and his recognition of the need for a horizontal rhetoric, gradual and indirect, an inexorable building of perception, bewilderment, anxiety, and curiosity. The two senses of wonderment create a highly tensive state out of which the physician, who has come "upon the knowledge that he has done no more than meddle in the lives of his fellows, and that he has done at least as much harm as good . . . may continue to pretend, at least, that there is nothing to fear, that death will not come, so long as people depend on his authority. Later, after his patients have left, he may closet himself in his darkened office, sweating and afraid" (Selzer 1976, 18). And the reader realizes that Selzer's essay itself has perplexed and bewildered and offered visions of horrible deeds, destruction, distress, and grief. He has made the reader a party to the amputation and created a state of mind much like that of the unbelieving doctor. The reader, too, is "sweating and afraid" and has reached the same "wonderment" by having learned that disease and disfigurement will come and that the authority of medicine's hierarchical rhetoric cannot control, prevent, or even soften such mortal lessons. Should the surgeon and the reader not find a means of transcending the stasis of such a wonderment, they may find themselves living the terrible life of Miguel de Unamuno's Don Immanuel.

In Unamuno's story, Don Immanuel is a beloved priest. In the eyes of the people of his village he "is already a saint," and he "transports them to paradise with his chanting." But, "the fact is that Don Immanuel is not so much a saint as a martyr. Long ago his own faith left him. He is an atheist, a good man doomed to suffer the life of a hypocrite, pretending to a faith he does not have. As he raises the chalice of wine, his hands tremble, and a cold sweat pours from him. He cannot stop for he knows that the people need this of him, that their need is greater than his sacrifice" (Selzer 1976, 18–19). Though he may not believe, Don Immanuel, like Selzer's unbelieving doctor, cannot simply walk away. His chants, rituals, and ministrations may be empty of the substance of his belief, but they create, for the people who depend upon his "authority," a powerful symbolic reality that makes life bearable, offering them not truth but hope. In his speech notes, Selzer describes the situation of the surgeon in a similar predicament:

> The . . . surgeon has only an airy bag of abstractions. . . . He dwells on them, recites them: It is for the good of mankind; it is to relieve pain. But something in him cries out against his list: it is perverse to try to relieve pain by inflicting it. Requires the giving over of the patient's free will and his trust. It is too much to ask. Yet we do it every day, and with an arrogance born of habit and custom, and grown casual, even charming.

"Still . . . still . . . could it be that Don Immanuel's whole life is a kind of prayer, a paean to God?" (Selzer 1976, 19). Certainly this is one way of perceiving such a life. Certainly both priest and unbelieving surgeon do good for those they tend. But the surgeon and the priest who serve in such a way cannot transcend the facts of their lives and are in no way freed of their burdens of unbelief and guilt. They sacrifice themselves and become, like the biblical prophet Jonah, unwilling, unrepentant, miserable servants (Selzer 1985). Still, Selzer's aim is not to portray the wondering doctor as a helpless failure, as other writers have done.

That the surgeon could reach the "state of wonderment" at all suggests that he may also transcend its stasis. This transcendence is, of course, precisely what Selzer's entire essay has built toward, a transcendence through which the unbelieving doctor becomes a writing doctor, a poet who treats his patients not with authority but with reverence. Selzer's doctor-writer has realized the limitations of his science, of the purely technical, the cutting, excising, and stitching that make up the surgeon's day-to-day experience. He reaches past the "impacted masses of his guilt" into the "high state of wonderment," a state Emerson describes as the prerequisite of the poet who "hears a voice, . . . sees a beckoning. Then he is apprised, with wonder, what herds of daemons hem him in. He can no more rest. . . . He pursues a beauty, half seen, which flies before him. . . . it is of the last importance that these things get spoken" (Emerson 1960, 239). In Selzer's surgical theater, the feeling that produces such a need to speak becomes so great that it overwhelms the surgeon, suffocating him. "Like an asthmatic hungering for air, longing to take just one deep breath, the surgeon struggles not to feel. It is suffocating to press the feeling out" (Selzer 1976,

101). To become an Emersonian poet is to set oneself free, is to breathe deeply, "is like the stock of air, for our respiration . . . not a measure of gallons, but the entire atmosphere if wanted" (Emerson 1960, 240).

The breathing simile perfectly captures the release that entering the "high state of wonderment" brings about when its paralysis-producing feeling, surprise, and "self-doubt" become perceived relationships and new understanding through the power of metaphor (Ricoeur 1984). The change, in both rhetorical and metaphorical perspectives, is remarkable.

> A writing doctor would treat men and women with equal reverence. . . . Women are physics and chemistry. They are matter. It is their bodies that tell of the frailty of men. Men have not their cellular, enzymatic wisdom. Man is albuminoid, proteinaceous, laked pearl; woman is yolky, ovoid, rich. Both are exuberant bloody growths. I would use the defects and deformities of each for my sacred purpose of writing, for I know it is the marred and scarred and faulty that are subject to grace. I would seek the soul in the facts of animal economy and profligacy. (Selzer 1976, 19)

The language, still replete with unexpected metaphors and new imagery, is no longer unsettling and debilitating in the way it has been throughout the essay. It invites a re-creation of one's sense of what men and women might be, a re-creation that clarifies, softens, and balances the vision of rotting flesh and dismemberment that Selzer has already offered. And where the language shocks, the shock pushes the reader upward toward wisdom, exuberance, and grace, not downward into mortality and death.

Two allusions to "The Poet" in the passage above suggest how and why the transformation from stasis and despair to exuberance takes place. "Every new relation is a new word," Emerson says,

> we use defects and deformities to a sacred purpose, so expressing our sense that the evils of the world are such only to the evil eye. In the old mythology . . . defects are ascribed to divine natures, as lameness to Vulcan, blindness to Cupid, and the like—to signify exuberances. For as it is dislocation and detachment from the life of God that makes things ugly, the poet, who re-attaches things to nature and the Whole—re-attaching even artificial things and violation of nature, to nature, by a deeper insight—disposes very easily of the most disagreeable facts. (1960, 229)

While Selzer is certainly concerned with "the most disagreeable facts," there is an important difference between his and Emerson's senses of the poet's ultimate aim with regard to them. For Emerson, "defects and deformities" are the objects of the poet's transmutative powers, which, through perception, correspondence, metaphor, and publication, bring them into accord with a larger, coherent, more attractive vision of the creation. Emerson uses "dislocation" and "re-attachment" to signify the deviant nature of the disagreeable facts of "ugliness" and "the evils of the world." Furthermore, he holds that "all form is an effect of character; all condition, of the quality of life; all harmony, of health; and for this reason a perception of beauty should be sympathetic, or proper only to the good. . . . The soul makes the body" (1960, 227). That is, for Emerson, the outward form is an index of the inner soul. The more perfect the body, the more perfect the soul that "made" it, and vice versa.

Selzer's is a nature in which outward form has little to do with beauty, which is not a first condition from which the disagreeable has fallen and to which it must be "re-attached." It is a nature in which the poet, because he is also a surgeon bound by the facts—a surgeon who must "pay meticulous attention, give reverent homage to the material, the materia medica, the hardware, the machinery of surgery" (Selzer, speech notes)—cannot dispose "very easily of the most disagreeable facts." The surgeon-poet's job is to perceive and to transmit the ancient beauty that arises from the interaction of the human beings he tends and the disagreeable facts of his surgical profession, not seeing ugliness or the evils of the world in terms of a more pleasingly formed beauty, but seeing all three in terms of the human spirit and revealing them to be integral parts of the total human experience.

For Selzer, the poet's power lies in his ability to offer his readers a verbal reality in which such painful mortal lessons as the diabetic woman's amputation may generate a love with the power to bring isolated individuals into a human community that makes its wounded, crippled members whole through what Morris terms "understanding and vision, which can recognize in mute, inexplicable pain the possibilities of knowledge, healing, and beauty" (1980, 128). To recognize such beauty is to transcend the defects and deformities of the body diseased, not to transmute them.

Although the work of Selzer's surgeon-poet has an aim quite different from Emerson's "true doctor," the means the two would employ and their relation to the phenomena about which they write are very much the same:

The poet turns the world to glass, and shows us all things in their right series and procession. For through [his] better perception he stands one step nearer to things, . . . [perceiving] that thought is multiform; that within the form of every creature is a force impelling it to ascend to a higher form. . . . All the facts of the animal economy, sex, nutriment, gestation, birth, growth, are symbols of the passage of the world into the soul of man, to suffer there a change and reappear a new and higher fact. (Emerson 1960, 230–31)

The twin forces that impel Selzer's patients, his physicians, and his readers to such a "higher form" are identification and love. These forces emerge from and are driven by "defects and deformities," conditions that make men and women "subject to grace," a word that means both "mercy" and a "seemingly effortless beauty" (*American Heritage Dictionary* 1971), a word referring, again, not to "a quality of form, but a quality of vision: a way of seeing, of feeling, of understanding" (Morris 1980, 127). "Yes," Selzer writes, "it is the exact location of the soul that I am after. The smell of it is in my nostrils. I have caught glimpses of it in the body diseased. If only I could tell it. Is there no mathematical equation that can guide me? So much pain and pus equals so much truth?" (Selzer 1976, 19).

Selzer goes on to say that once he thought he had "it," and tells a second "representative anecdote," this one representative of the "normal" reality of the surgeon deep in the "love affair with his technique." The language of this anecdote is a dialect of scientific medicine's language, and it operates according to the most extreme rules of the hierarchical rhetoric that dominates and drives that language, rules whose discursive underpinnings arise from metaphors of war, imprisonment, hatred, and death. The story is about a young surgeon who, through his technical expertise, his surgical science, and the extirpative power of his instruments, would save the world, a young surgeon who thought he had found the "equation."

A young man, recently returned from an archaeological expedition to Guatemala, comes to the clinic in which the surgeon, exhausted after a long day's work, labors. The young man has on his upper arm what the surgeon is certain is an abscess. The surgeon says, "I will enlarge the opening to allow better egress of the pus." But it is not an abscess, for out of the "crater" there "emerges a narrow gray head whose sole distinguishing feature is a pair of black pincers. The head sits atop a longish flexible neck arching now this way, now that, testing the air. Alternately it folds back upon itself, then advances in new boldness. And all the while, with dreadful rhythmicity, the unspeakable pincers open and close" (1976, 20).

In the face of such an outrageous complication the surgeon is astonished, stunned by this "beast at whose malignant purpose I could but guess." "With all the ritual deliberation of a high priest I advance a surgical clamp toward the hole," but the clamp misses the "Mayan Devil." The surgeon waits ("One must skulk"), and then undergoes a transformation. "Acrouch, strung, the surgeon is one with his instrument; there is no longer any boundary between its metal and his flesh. They are joined in a single perfect tool of extirpation. It is just for this that he was born. Now— thrust—and clamp—and yes. Got him!" The surgeon feels the creature struggle in his grasp, hears its "brittle scream," and is seized with a hatred, "such a detestation as would make of Iago a drooling sucktit. It is the demented hatred of the victor for the vanquished, the warden for his prisoner. It is the hatred of fear. Within the jaws of my hemostat is the whole of the evil of the world, the dark concentrate itself, and I shall kill it. For mankind. And, in so doing, will open the way into a thousand years of perfect peace. Here is Surgeon as Savior indeed" (1976, 19–20).

Here is surgeon as defined by the normal codes of his profession, "a perfect tool of extirpation," at one not with his patients or humanity but with his instrument, acting for the good of all by killing the "evil of the world"; in Emerson's terms, disposing "very easily of the most disagreeable facts." What startles most is the astonishing number of divisions that form the context within which this "oneness" happens and the violence with which those divisions are maintained: pincers open and close, the head advances and retreats, the two "jaws" of the hemostat "clamp," the victor and the vanquished, the prisoner and the warden, the one who hates and the thing that is hated, good and evil, dark and light, and the "Mayan Devil" and the "Surgeon as Savior." These are not divisions the young surgeon is likely to resolve because they are divisions built into the very language by which he creates what Benjamin Lee Whorf calls the "house of consciousness" within which he lives and works (1982, 213, 252). They are constitutive of his conception of perfection.

This young surgeon's "oneness" stands in marked and shocking contrast to the surgeon's relationship with his instruments at the blind diabetic woman's amputation. There, the saw seemed to cut by itself, to do its terrible, necessary work alone while the surgeon and the reader listened for "a little crack at the end." Their oneness came not from the manipulation of surgical instruments but from participation in the woman's human predicament. This participation led both the surgeon and the reader to perceive the beauty of the woman's response to the misery of her ulcer and to the pain of her surgeon. And finally, it afforded them all a glimpse of the most important human and rhetorical unity, "a love that blooms in the stoniest desert."

The young surgeon in this second story, because he understands himself to be "a perfect tool of extirpation," is turned, by the limitations of his language and its divisive, hierarchical rhetoric, away from love to hatred, detestation, and fear—emotions, in Ricoeur's sense of the word, that lead not to clarification or transcendence but to the "demented" responses of victor and warden (Ricoeur 1978). These responses shape his vision of the equation that would guide him to the "exact location of the soul," into the figure of a "Mayan Devil . . . with horrid blanket-wings and iridescent scales, raking, pincing, injecting God knows what acid juice," which he can only "solve" by tearing the creature from the body of his patient and then dropping it with complete assurance, "quickly . . . into the specimen jar of saline . . . the lid screwed tight. Crazily he swims round and round, wiping his slimy head against the glass, then slowly sinks to the bottom, the mass of hooks in frantic agonal wave" (1976, 21). Instead of reverence, this surgeon treats with authority, taking upon himself and his instruments the outlandish task of saving the world. " 'You are going to be all right,' I say to my patient. 'We are all going to be all right from now on' " (1976, 21). Such a pretentious authority is not acceptable to Selzer, and he is quick to deflate this "Surgeon as Savior indeed."

The following day, the surgeon takes the remains of the creature to the medical school for identification. " 'That's the larva of the botfly,' says a pathologist. 'The fly usually bites a cow and deposits its eggs beneath the skin. There, the egg develops into the larval form which, when ready, burrows its way to the outside through the hide and falls to the ground. In time it matures into a full-grown botfly. This one happened to bite a man. It was about to come out on its own, and, of course, it would have died' " (1976, 21–23). The surgeon is crushed. He knows "the mallet-blow of glory pulverized," and he learns, or rather the narrator asserts, that "it is not the surgeon who is God's darling. He is the victim of vanity. It is the poet who heals with his words, staunches the flow of blood, stills the rattling breath, applies poultice to the scalded flesh. Do you ask me why a surgeon writes? I think it is because I wish to be a doctor" (1976, 23).

In Selzer's case, the act of writing is immediate, literal, and necessary, but *writing* and *poet,* like so many other terms in Selzer's discourse, are not limited to single, literal senses. It is the transcendent, healing quality of the poet's vision and understanding that is crucial to doctoring, not the publication of books, stories, and essays. And this is the point of "The Exact Location of the Soul." The technical surgeon, intent on simply disposing of "the evils of the world," has yet to learn what the poet can teach—that nature is complex and multiform; that even those parts which seem terrifying and "dread-full" have a cause and purpose that elude his simplistic, violent technique; that health arises from wholeness; and most important, that healing flows both from and to patients and physicians.

REFERENCES

The American Heritage Dictionary of the English Language. 1971. Boston: American Heritage Publishing and Houghton Mifflin.

Burke, K. 1969. *A Rhetoric of Motives.* Berkeley: Univ. of California Press.

———. 1974. *A Grammar of Motives.* Berkeley: Univ. of California Press.

The Compact Edition of the Oxford English Dictionary. 1973. New York: Oxford Univ. Press.

Emerson, R. W. 1960. *Selections from Ralph Waldo Emerson: An Organic Anthology.* Ed. S. E. Whicher. Boston: Houghton Mifflin.

Morgan, T. 1979. Is There a Doctor in the House? *Saturday Review,* 6 August.

Morris, D. B. 1980. Beauty and Pain: Notes on the Art of Richard Selzer. *Iowa Review* 11: 124–30.

Ricoeur, P. 1978. The Metaphorical Process as Cognition, Imagination, and Feeling. *Critical Inquiry* 5(1): 143–60.

———. 1984. *The Rule of Metaphor: Multi-Disciplinary Studies in the Creation of Meaning in Language.* Trans. R. Czerny, with K. McLaughlin and J. Costello. Toronto: Univ. of Toronto Press.

Schuster, C. I. 1984. Confessions of a Writer: The Art of Richard Selzer. *Rhetoric Review* 3(September): 84–88.

Selzer, R. 1976. *Mortal Lessons: Notes on the Art of Surgery.* New York: Simon and Schuster.

———. 1983. Address to the College of Architecture at Cooper Union University, New York.

———. 1984. Interview with author. New Haven, Conn., 1–2 April.

———. 1985. Jonah and the Whale. *Medical Heritage* 1(1): 20–23.

———. n.d. Address to faculty and students at the University of Dallas, Texas.

———. n.d. Speech notes. These are unordered, undated, and unpaged notes from which Selzer draws material to compose his various public addresses.

Trautmann, J., ed. 1981. *Healing Arts in Dialogue: Medicine and Literature*. Medical Humanities Series. Carbondale: Southern Illinois Univ. Press, in cooperation with the Institute on Human Values in Medicine of the Society for Health and Human Values.

Whorf, B. L. 1982. *Language, Thought, and Reality: Selected Writings of Benjamin Lee Whorf*. Ed. J. B. Carroll. Cambridge: MIT Press.

Oliver Sacks's Neurology of Identity

MURDO WILLIAM MCRAE

It is impossible not to respect and admire Oliver Sacks, successful neurologist and best-selling author of five books, one of them the basis for Penny Marshall's popular 1990 film *Awakenings*. Once himself a victim of neurological impairment, Sacks displays immense sympathy and compassion for his patients, many of whom have endured lives of unimaginable suffering. Stylistically and rhetorically vigorous, even audacious at times, his philosophically sophisticated prose contains a virtually encyclopedic range of allusions, not only to neurologists such as Henry Head or Hughlings Jackson, but also to philosophers and poets such as Gottfried Wilhelm von Leibniz, Pythagoras and René Descartes, John Donne and W. H. Auden. Perhaps most appealing, however, is his passionate rejection of the mechanistic and depersonalized ways in which traditional neurology views each patient. Sacks advocates instead a "neurology of identity," superior to the neurological tradition because it would not regard each patient as a realization of some abstract neurological model (1985, xiv).

His conviction appears in *Awakenings* (1974), his first effort to supplant the neurological tradition, when he writes that "one must cease to regard all patients as replicas, and honor each one with individual attention, attention to how *he* is doing, to *his* individual reactions and propensities" (219). This stress on each patient's irreplicable individuality would bring Sacks eleven years later, in *The Man Who Mistook His Wife for a Hat* (1985), to contemplate a certain Dr. P., a musician who was unable to recognize concrete facial details as anything other than abstract, formal patterns. Sacks demonstrates his inclination to the use of tropes when he considers the significance of Dr. P., whose disorder is "an awful analogy . . . a warning and parable—of what happens to a science which eschews the judgmental, the particular, the personal" (19).

Since our culture locates human identity in the individual and unique, the irreplicable and particular, Sacks's various claims for the neurology of identity carry a powerful appeal. In spite of this, however, Sacks's repudiation of abstract and impersonal thinking should not blind us to the equally abstract and impersonal character of his own thinking. In the end, Sacks views his own patients, even himself when he is the patient, in ways that duplicate what he declares to be the conceptual, even ethical, fault in traditional neurology—its tendency to turn each patient into a replica of every other one.

One way to explain this duplication is to examine what Sacks describes as the Leibnizian and Pythagorean dimensions of his thought, both of which lead him to prefer the abstract over the concrete, thus to see the single empirical instance as a synecdoche for all other instances and for the form they realize. Since this preference means that Sacks thinks as traditional neurologists do, we might condemn him for being no less conceptually and ethically bankrupt than the tradition he scorns. That condemnation would be unfair, however, for Sacks cannot choose any other way to regard the patient. An explanation for why he cannot may be found in what Mikhail Bakhtin and Michel Serres have to say about *dialogue,* their term for the social character of language. Their comments are useful because they imply that scientific arguments are governed by the tendency, perhaps even the imperative, of regarding each concrete and particular datum tropically, as if it were a synecdoche for all the data, thus for the form they realize.

LEIBNIZIAN AND PYTHAGOREAN METAPHYSICS

Sacks indicts the neurological tradition by asserting that its founders—Henry Head, Hughlings Jackson, and C. S. Sherrington—always held to the false conviction that health "can be reduced to certain 'factors' or 'elements' . . . reduced to a *level,* something . . . titrated or topped-up in a mechanical way" (1974, 21). To this "Newtonian-Lockean-Cartesian view" (1974, 191) Sacks brings the perspective of Leibniz, whose God knows how every body in the universe "feels the effect of all that takes place" and thus can "read in each what is happening everywhere" (Leibniz 1948, 251). For Sacks, then, "health is infinite and expansive in mode, and reaches out to be filled with the fullness of the world" (1974, 196). This rejection of Cartesian mathematics, Newtonian physics, and Lockean psychology also means that a reformed neurology should try to comprehend each patient in terms of what Sacks calls Leibnizian "metaphysics," which "makes no such [mechanistic, Newtonian-Lockean-Cartesian] reductions: its terms are those of organization or design" (1974, 21).

True enough, Sacks's critique of traditional neurology makes no use of strictly mechanistic paradigms.[1] But consider, for example, *The Man Who Mistook His Wife for a Hat*, in which Sacks organizes the study of neurological disease by dividing it into four abstract categories: "losses," "excesses," "transports," and the "world of the simple." There is something attractive, perhaps, in the way each category points to what Sacks claims to be an epistemological failing in traditional neurology. Thus, "losses" inquires into why the tradition can think of patients only in terms of their functional deficits; "excesses" and "transports" attempt, among other things, to explain why the tradition can never regard disease, paradoxically, as a sort of health; and "the world of the simple" examines how the mentally disabled often possess intellectual skills far in excess of those possessed by the mentally able.

Interesting though they are, however, the abstract character of these categories self-evidently betrays Sacks's emphasis on the unique. That is, the key terms of his metaphysics—*organization* and *design*—recapitulate the neurological tradition's way of thinking. Implicitly comparing his models with the tradition's, Sacks encourages traditional neurology to displace its mechanistic sense of organization and design with his universalizing one. Such a comparison sharpens his ability to characterize the tradition as reductive, but to make that comparison forceful he must also conversely presume that his competing notions of organization or design are more encompassing than those that shape the neurological tradition. In this respect, Sacks's move from the supposedly empirical reductions of Isaac Newton, John Locke, and Descartes to the expansive metaphysics of Leibniz is not unexpected.[2] Nonetheless, when he speaks out against reductive neurology, his elevation of metaphysics over reductiveness reverses the valuation of the particular over the abstract evident in his tropical description of Dr. P.

In other words, if thinking of each patient as a replica of the others is truly "to commit an epistemological solecism of the first order" (1974, 202), then the solecism is Sacks's as well, given his metaphysics of organization or design. Patterns displace particulars in Sacks's thought, which explains why he confesses to a lifelong Pythagorean enchantment with the formal properties of numbers. In *A Leg to Stand On*, his account of treatment for nerves severed in his left leg during a hiking accident, he describes the most formative intellectual experience of his childhood, when, in his family garden, he would gaze upon the sunflowers,

> whose vast inflorescences fascinated me endlessly and showed me at five the Pythagorean mystery of the world. (For it was then, in the summer of 1938, that I discovered that the whorled florets were multiples of prime numbers, and I had such a vision of the order and beauty of the world as was to be a prototype of every scientific wonder and joy I was later to experience.) (1984, 34)

The abstract and the computational are crucial for Sacks's life as a scientist, but this Wordsworthian memory also recalls how Sacks used the description of Dr. P., the analogy of a science that "becomes entirely abstract and computational" (1985, 19), to condemn the neurological tradition.

Given its universalizing power, the Pythagorean remains constant in Sacks's thought, appearing again in *The Man Who Mistook His Wife for a Hat*, in which he describes two clinical patients, John and Michael, twin autistic savants who possessed a " 'Pythagorean' sensibility" of recognizing almost instantly, with no computational aids, any prime number upward of twenty digits (1985, 197). Their skill so reminded Sacks of himself as a childhood "number brooder" that he decided to locate a table of prime numbers in order to participate in their "numerical communion" (1985, 192), their conversing with each other by stating little more than prime numbers. Eventually Sacks was able to hold a sort of conversation with them, one whose sole content was his stating of a prime number and the twins' responding to him by stating an even larger one.

Although his account of this conversation possesses no small measure of charm—the doctor sprawled on the floor, entering with "great joy" (1985, 193) into the twins' number game—the foundation of that conversation was Sacks's reacting to the twins on the level of their abstract computational skills. The twins repeatedly claimed that they mentally "saw" (1985, 189) each prime number in their computational games, but what Sacks perceived was how one twin replicated the pleasure the other experienced when recognizing any prime. At the moment either twin saw any given prime, Sacks further regarded him to be intuitively and joyfully perceiving the pattern of all primes (numbers divisible only by themselves or by one).

Just as in his family's garden Sacks had seen the empirical part instantly disappear into the formal whole—the florets *were* multiples of prime numbers—so here, in conversation with the twins, what he saw was not a John or a Michael but a living prime number which said itself to another prime

number in the replicable pattern of all primes. Replicating the other's computational skills, each twin for Sacks thus did not "merely live in a world
of numbers, but in a world, in *the* world, *as* numbers" (1985, 203). Each
twin finally occupied in Sacks's thought the place of a living number from a
table of prime numbers, a part that realizes the formal pattern of the whole.

As these several examples indicate, Sacks often does conceive of the
individual patient as a replica of other patients, their suffering evoking
the model of their disease. No wonder that, in *Awakenings*, his account
of treating postencephalitic parkinsonism with the drug L-DOPA, he describes each of his patients as if their disease totally absorbed their being:
"from its first, infinitesimal, intimation or twinge, [the disease] could pro-
ceed by an infinite multitude of infinitesimal increments to an infinite, and
then more infinite, and still more infinite, degree of severity . . . its 'least
part,' so to speak, possessed (in infinitesimal form) the entire, indivisible
nature of the whole" (1974, 78, n. 18). Making a universalizing, Leibnizian
claim that the indivisible whole is realized in even the most infinitesimal
instance, this passage turns on the infinitesimal standing in for the whole,
in precisely the relationship of part to whole that defines the most common
instances of synecdoche.[3] This sort of thinking ultimately brings Sacks to
insist that "one's total being, one's relation to the world, is in a sense present
and represented in each and every part of the brain (or, at least, the cortex);
and thus, that one's ontological organization is essentially *microcosmic* or
monadic in kind" (1974, 261, n. 11).

These ontological claims in *Awakenings* join with assertions concerning
the triadic sequence of "awakening, tribulation, accommodation," which
constitutes the metaphysical organization or design realized in the history of each victim of the disease. "It is in terms of this sequence," Sacks
argues, "that we can best discuss the consequences of L-DOPA" (1974,
196–97): the way that the drug first euphorically awakens the postencephalitic patient, eliminating his stupors, tics, and overwhelming mus-
cular disorders, only to plunge him eventually into even more profound
parkinsonian tribulations, from which there is no further release, only
accommodation to the disease.

Through this sequence, Leibnizian universalism explicitly merges with
the tropical in Sacks's thought, leading him to claim that other "metaphorical triads" or "allegorical sequences" could be used to characterize
the progress of L-DOPA treatment: for example, the triads of "peace,
war, truce"; or "first bliss, bliss lost, bliss regained"; or "at-homeness,
departure, return" (1974, 197, n. 8). Sacks thus implies that the sequence

"awakening, tribulation, accommodation" is part of some larger whole named by a sequence such as "peace, war, truce" or "at-homeness, departure, return." We should recognize, however, that the reverse could also be maintained. Sequences such as "peace, war, truce" could be subsumed, for example, in the metaphorical triad "awakening, tribulation, accommodation." This slipperiness of parts and wholes illustrates both how the whole can take the place of the part and how thoughts of the formal and not the empirical ultimately govern the operation of synecdoche. Indeed, this is precisely why all of the sequences Sacks adduces ultimately realize an even more abstract model—"the course of all our lives, and of history and culture" (1974, 197, n. 8).

I draw attention to Sacks's habits of thought not to quibble with him, and certainly not to gainsay his compassion, but to respond to the explicitly philosophical and rhetorical character of his arguments. Since his forceful advocacy of the neurology of identity rests on claims regarding how best to regard the individual patient, it is appropriate to ask whether he regards his own patients in terms of those claims. His Pythagorean and Leibnizian vision of the patient as one who constitutes the same organization or design as every other patient self-evidently moves Sacks some distance from claims for each patient's irreplicable uniqueness. Granted, Sacks must conceive of disease and human existence in these terms, for without them he could not articulate his personal understanding of organization or design. Even so, precisely because he conceives the patient to be monadic and not atomistic, interconnected with all others and not radically individuated, Sacks always insists that "the study of identity and disease cannot be disjoined" (1985, xiv): the patient's individual identity resides not in his particularity but in the abstract form of his disease.[4] Instead of inappropriately charging Sacks with bad faith, however, we need to examine why he cannot choose some other way to regard the patient.

SERRES, BAKHTIN, AND THE DIALOGIC

When Sacks decries reductive and abstract thinking in the neurological tradition, he must envision himself engaging the tradition in a dialogue whose aim is to reform the way it conceives the patient. When we look at Sacks as a participant in such a dialogue, however, we see that little separates his way from the traditional way of thinking. To account for why this is ineluctably the case, we need to bring both Mikhail Bakhtin and Michel

Serres into the conversation, for under the notion of the dialogic, both have thought deeply about the connection between language and rhetoric, on the one hand, and about single instances and formal wholes, on the other.

When Bakhtin thinks about dialogue, he emphasizes how the social dimensions of language guarantee individuality. Like Sacks, Bakhtin opposes the reductive; in his case, the reductions of the domineering, or monologic, "author who finds it easy to purge his work of speech diversity: he simply does not listen to the fundamental heteroglossia [diversity] inherent in actual language; he mistakes social overtones, which create the timbres of words, for irritating noises" (1981, 327). "A living human being," Bakhtin consequently insists, "cannot be turned into the voiceless object of some secondhand, finalizing cognitive process. *In a human being there is always something that . . . does not submit to an externalizing secondhand definition*" (1984, 58).

When Sacks argues for uniqueness, he sounds much the same as Bakhtin, who tells us to pay attention to speech diversity in order to understand how identity cannot be submitted to an externalizing, secondhand definition. In fact, when Sacks first proposes the term *neurology of identity*, he expresses his own disdain for the typical medical case history, a kind of monologizing, externalizing, and secondhand definition of a human being, and advocates instead what might be called his own dialogic narrative principle: "we must deepen a case history to a narrative or tale" (1985, xiv) of individual suffering. Sacks's obedience to this principle, after all, seems to explain the way he organized both *Awakenings* and *The Man Who Mistook His Wife for a Hat* as collections of narratives about individual patients.

Unlike Sacks, however, Bakhtin does not imply that identity emerges in opposition to the collective. True enough, Bakhtin writes that a speaker's "word encounters an alien word and cannot help encountering it in a living, tension-filled interaction" (1981, 279), but this interaction does not mean that each verbal exchange pits the individual against the social. On the contrary, any word a speaker utters is so charged with social overtones that it "is half someone else's" (1981, 293). Speech marks no purely subjective, individual experience for Bakhtin; the "individual is a purely socioideological phenomenon . . . [and] the content of the 'individual' psyche is by its very nature just as social as is ideology," he writes under Vološinov's name; "I give myself verbal shape from another's point of view, ultimately, from the point of view of the community to which I belong" (1973, 86).[6] The social languages of a speaker's age group, social level, academic institution, profession, and even family are "forms for conceptualizing the world in words, specific world views" (1981, 291–92). Speakers utter nothing that is not thoroughly social, their expression in every respect at the moment of its utterance a particular reiteration of a socioideological point of view.

This is the point implied in an amusing, if delicately phrased, passage from Dostoevsky's *Diary of a Writer*, which Bakhtin cites at length in order to comment on "expressive intonation," or the way connotation often exceeds denotation (1973, 103–4). Dostoevsky writes of one evening coming upon "six tipsy artisans" who seemed deeply engaged in a trivial, yet heated, debate. What made that debate so curious, however, was that the only word spoken was an "extremely laconic designation of a certain item, and nothing more"; that is, writes Dostoevsky, the indelicate reiteration of a "noun whose usage is forbidden in the company of ladies." Twice Dostoevsky draws attention to this single word—"a certain noun, a noun, moreover, of utmost simplicity," he says at one point; "just that one single word, just that one word alone," he writes at another—in order twice to emphasize how "whole trains of reasoning," a "whole train of thought," can be evoked in but a single word.

Now, if it seems that the obscene conversation of six drunken Russians has no direct bearing on the ways Oliver Sacks conceives of his patients, recall, for example, the dialogue Sacks had with the autistic twins John and Michael; how he was able to communicate with them only when he regarded them as synecdoches for the formal pattern of all prime numbers. In the same way that John and Michael's conversation reveals their numerical "point of view," so the single obscenity exchanged by Dostoevsky's interlocutors reveals their socioideological points of view. Indeed, Bakhtin understands that Dostoevsky could not recognize how each reiteration of the obscenity stood in for a point of view unless each utterance somehow recalled other expressions of that point of view. Similarly, Sacks could not conclude that the twins conversed in prime numbers unless he recognized how each number they spoke evoked the pattern of prime numbers.

Dostoevsky's story illustrates for Bakhtin how a single statement evokes a collective point of view, but not that any utterance expresses every point of view in "the Tower-of-Babel mixing of languages" (1981, 278) that is social dialogue. A speaker's utterance must remain selective to be heard; expression must exclude certain nuances of words, certain socioideological points of view, and yet include others. This is why Bakhtin conceives of dialogue as does Michel Serres, who writes that even when participants in a dialogue seem to disagree, "interlocutors are in no way opposed . . . they are on the same side, tied together by a mutual interest: they battle together against noise" (1982a, 67). And this is also why, though Sacks calls the neurological tradition into question, we sense how he is more fundamentally its dialogic ally, working with it against the "demon," "the third man," "the parasite"—all terms that Serres (1982b) uses elsewhere to define any mass of formless observations, any "noise," which must be silenced if any dialogic, formal abstraction is to occur. Both Sacks and the tradition he condemns, yet therefore speaks to in a sort of dialogue, make use of an identical strategy for conceiving the patient: in Serres's terms, both "eliminate the empirical," the individual and unique patient, and thereby "make a science possible" (1982a, 69).

Serres would insist that both Sacks and the tradition hope to format into useful data what would otherwise remain a mass of noisy, formless observations about patients. Sacks wants the tradition to realize a different form in the empirical, but not to ignore the formal altogether, since, as Serres puts it, the "first effort to make communication in a dialogue successful is isomorphic to the effort to render a form independent of its empirical realizations" (1982a, 69).[7] In a sense, Sacks wants the tradition to allow the facts about patients to speak for themselves, but this can only occur when the empirical gives way to the formal. Only then do the facts seem to speak for themselves; only then may they be heard against the noise, which ceases to sound at the precise moment the facts take on their formal character. This is the very foundation of the concept of form, and of any effort to abstract the form from the data. When Sacks speaks against reductive thought, in other words, he is actively inviting the tradition to hear what he hears, to silence the noise by formalizing it.

Were Serres to read the works of Oliver Sacks, he would no doubt explain that Sacks's tendency to formalize is far from being an epistemological solecism, but is instead so essential that even when Sacks regards himself as a patient, as he does in A Leg to Stand On, he gazes upon himself in a formalizing way.[8] Passionate and moving though it is, Sacks's narrative of his intense personal suffering perfectly illustrates Serres's point that to communicate is to "eliminate that which hides form" (1982a, 69), to render the form independent of its individual empirical realizations. The symmetry between dialogue and the discovery of form explains why the story of Sacks's treatment turns on his feeling excommunicated, both from his left leg, a lump of flesh that would not respond to his call, and from his physicians, who repeatedly assured him that he was unique.

What is initially remarkable about Sacks's narrative is the way he tells us that his left leg, whose nerves had been severed in a hiking accident, refused to communicate: "When I called to the muscle, there was no answer. My call was not heard, the muscle was deaf" (1984, 66). The silence of his uncommunicative leg meant he no longer could regard himself as a being unified through proprioception, or the way "the body knows itself, and has itself as 'property' . . . by virtue of a constant flow of incoming information, arising ceaselessly, throughout life, from the muscles, joints and tendons" (1984, 71). So profound was the tear in his being created by his uncommunicative leg that it left Sacks feeling that part of the "inner photograph" (1984, 75) of his body was missing, that the image of his body as a complete and unified whole had vanished.

Disturbing though it was, the proprioceptive terror that overcame Sacks in those awful days after his accident occurred because he defined proprioception as a sort of bodily colloquy that discovers a form—in this case, the unified body image. That unified image must extend even to the most infinitesimal part of the body if it can be lost in the uncommunicative, nonproprioceptive absence of any part. Sacks's response to his leg recalls the Leibnizian manner in which he regarded his L-DOPA patients, sensing their parkinsonism to extend to the least part of their being. Accordingly, when sensation returned to his leg at a moment when Mendelssohn's music unexpectedly came to mind, his sense of a complete and unified being, an "I" who was the expression of an organizing pattern, also returned, with equally Leibnizian force: "what appeared in this moment transcended the physical, but instantly organized and reorganized it into a seamless perfect Whole. . . . Grace, unbidden, appeared on the scene . . . and instantly coordinated, subordinated, all phenomena to itself" (1984, 150).

This moment of recovery suggests that Sacks could very well have delineated another triadic sequence—call it "communication, excommunication, communion"—to formalize his own experiences as a patient. His body's proprioceptive communication was disrupted by the excommunicative silence of his injured leg, but health began to return when his leg experienced a sort of communion with the bodily *plenum*. This triad could also join the others listed in *Awakenings*, as a way of stressing how Sacks's own suffering recapitulates the history of other patients: the twins John and Michael in their "numerical communion"; or the memory-impaired alcoholic, Jimmy G., who could nonetheless experience the "fullness and totality of Communion, the perfect alignment of his spirit and the spirit of the Mass" (1974, 36); or the parkinsonian Martin, for whom "communion with the music" of the church choir brought "absorption and animation, wholeness and health" (1974, 183); or even Sacks himself in the convalescent home, where sitting one September morning on a stone bench, like "Adam beholding the new world with wonder," he felt "a perfect peace and communion lay upon the land" (1984, 174).

Just as Sacks's relationship with his nonproprioceptive leg was a failed dialogue, so also were his contacts with his physicians. When he attempted to explain his loss of body image, his surgeon felt that there was nothing to be alarmed about. Sacks writes that his surgeon "didn't even listen to me. He showed no concern. He doesn't listen to his patients— he doesn't give a damn" (1984, 105). Later, his surgeon's senior registrar spoke with him, assuring him, "Sacks you're unique. . . . I've never heard anything like this from a patient before" (1984, 106). But those assurances provoked Sacks's rage, a reaction quite unexpected from a neurologist who wants to institute the neurology of identity: "'*I can't be unique,*' I said, with anger, and rising panic. '*I must be constituted the same way as everyone else!* . . . you don't listen to what patients say, perhaps you're not interested in the experiences they have'" (1984, 106–7; emphasis added).

Sacks's convalescence further extended his feeling of being constituted no differently from other neurologically impaired patients. In convalescence, he learned that "my own experience, my 'case,' was *far from unique.* Almost every patient who had injury or surgery to a limb, and whose limb had then been casted, out of sight, out of action, had experienced some degree of alienation" (1984, 161; emphasis added). Again, there is no denying his great sympathy and compassion for his fellow patients, for he intimately recognized the terror and alienation they experienced, but it still remained for him that his fellow patients' identities resided not in their uniqueness but in the way each served as a synecdoche for an abstract form: the alienated, excommunicated, cast-out patient.

Throughout his arguments for a neurology of identity, Sacks faults a scientific way of thinking that is as fundamental to him as it is to the neurological tradition. This is why, regardless of his sympathy and compassion for patients, or the vigor of his style, or the intensity of his commitment to reform his profession, his efforts must be regarded with some caution, though not outright condemnation. Sacks's arguments repeatedly demonstrate that whenever the one instance is thought somehow to replicate the others (and this is fundamental to the Leibnizian and Pythagorean character of his intellect), there is consequently a tendency to regard each instance tropically, as a synecdoche for all other instances, and thus for the form, organization, or design they realize. To accept Sacks's indictment of reductive and abstract thinking without asking whether his own conceptual mechanics is fundamentally different would be to ignore how empirical science tries to understand each datum as an instance of an abstract form replicated in all the data.[9] In the end, we may respect and admire his efforts, and we cannot condemn him out of hand, but Oliver Sacks never understands what Serres and Bakhtin implicitly do: synecdoche directs the traffic at the crossroads of science, rhetoric, and epistemology.

NOTES

1. Elsewhere in this volume, Mary Ellen Pitts analyzes Sacks's rejection of mechanistic thought. I agree with Pitts that Sacks values the kinesthetic over the mechanistic; doing so is the source of his compassion for patients. Compassion does not necessarily guarantee the absence of abstract thinking, however.

2. In spite of his claims otherwise, there is an interesting symmetry between Sacks's ways of thinking and Newton's. In their survey of classical physics, Ilya Prigogine and Isabelle Stengers describe Newton's spatial dynamics in ways that could also characterize how Sacks maps his conceptual space: "any single state is sufficient to define the system completely. . . . Dynamics defines all states as equivalent: each of them allows all the others to be calculated" (1984, 60).

3. Synecdoche rather than metonymy names Sacks's dominant trope because his conception of the patient's status relies less on the metonymic "changing" or "substituting" of names than on the synecdochic "receiving together" of characteristics. Such a distinction is in keeping with Roman Jakobson's seminal meditations on the metaphoric and the metonymic. In cubism, for example, Jakobson observes that "the object is transformed into a set of synecdoches" (1971, 1114). As any cubist figure is the empirical realization of certain formal, geometric patterns, it synecdochically calls up all other cubist figures. Jakobson thus implies my point that synecdoche is the trope that realizes the formal in the empirical.

4. Elsewhere in Sacks, the fascination with abstract organization and design first surfaced in *Migraine: Understanding a Common Disorder* (1972, rev. 1985), in which his goal was "to trace the basic design or *structure* of migraine," especially its five stages of "excitement," "engorgement," "prostration," "resolution," and "rebound" (120). The same fascination appears in *Seeing Voices: A Journey into the World of the Deaf* (1989), his most recent book, a study of the sign languages of the deaf, in which he writes that "the beauty of . . . [American] Sign . . . is like the beauty of theory in this way: that the concrete leads to the general, but it is through the general that one recaptures the concrete, intensified, transfigured" (123).

5. In his thematic survey of Sacks's oeuvre, William Howarth remarks on the "classical" and "romantic" in Sacks's writing to show how Sacks develops nar-

rative forms that "establish a full ecology of healing, a view that emphasizes dynamic exchange, recurring cycles, the balance and integrated harmony of health" (1990, 119).

6. For an argument ascribing Vološinov's text to Bakhtin, see Holquist (1981).

7. For Serres, mathematics constitutes the most pronounced expression of this isomorphism of dialogue and scientific abstraction, for it is "an ideal republic which is the city of communication maximally purged of noise" (1982a, 68). No wonder, then, the Pythagorean in Sacks's thought, the way it conceives the autistic John and Michael, for example, to be synecdoches for the formal pattern of prime numbers.

8. I refer to Sacks's "gaze" in order to borrow from Michel Foucault's history of how the nineteenth-century birth of the clinical "gaze" reveals a tendency to regard the patient as a synecdoche. Like Serres, Foucault attends to the relationship between the formal and the empirical, on the one hand, and to the expression of that relationship, on the other. In the clinics of the nineteenth century, "the individual in question was not so much a sick person as the endlessly reproducible pathological fact to be found in all patients suffering in a similar way," Foucault observes, and "to the exhaustive presence of the disease in its symptoms corresponds the unobstructed transparency of the pathological being with the syntax of a descriptive language: a fundamental isomorphism of the structure of disease and of the verbal form that circumscribes it" (1973, 97).

9. T. S. Kuhn also implies that a datum is a synecdoche for the form when he describes how "exemplars," or a scientific "community's standard examples," enable its members to acquire the community's "disciplinary matrix" (1977, 293–319). Kuhn adopted the terms *exemplar* and *disciplinary matrix* in response to criticism that he deployed the term *paradigm* in his *Structure of Scientific Revolutions* (1962, rev. 1970) in seemingly contradictory ways. But since paradigm involves a sort of merging of the formal and the empirical, it seems to me to indicate precisely how the empirical replicates the formal. If there is slippage in Kuhn's use of *paradigm*, it does not mark contradiction; indeed, labeling Kuhn contradictory misses the supplementary nature of his case; see McRae (1988).

REFERENCES

Bakhtin, M. [V. N. Vološinov, pseud.] 1973. *Marxism and the Philosophy of Language*. Trans. L. Matejka and I. R. Titunik. New York: Seminar Press.
———. 1981. *The Dialogic Imagination: Four Essays*. Trans. C. Emerson and M. Holquist. Austin: Univ. of Texas Press.
———. 1984. *Problems of Dostoevsky's Poetics*. Trans. C. Emerson. Minneapolis: Univ. of Minnesota Press.
Foucault, M. 1973. *The Birth of the Clinic*. Trans. A. M. S. Smith. New York: Parthenon.

Holquist, M. 1981. The Politics of Representation. In *Allegory and Representation: Selected Papers from the English Institute, 1979–80*. Ed. S. J. Greenblatt. Baltimore: Johns Hopkins Univ. Press.

Howarth, W. 1990. Oliver Sacks: The Ecology of Writing Science. *Modern Language Studies* 20(4): 103–20.

Jakobson, R. 1971. The Metaphoric and Metonymic Poles. In *Critical Theory since Plato*. Ed. H. Adams. New York: Harcourt.

Kuhn, T. S. 1970. *The Structure of Scientific Revolutions*. 2d ed. Chicago: Univ. of Chicago Press.

———. 1977. *The Essential Tension: Studies in Scientific Tradition and Change*. Chicago: Univ. of Chicago Press.

Leibniz, G. W. von. 1948. *The Monadology and Other Philosophical Writings*. Trans. R. Latta. Oxford: Oxford Univ. Press.

McRae, M. W. 1988. The Paradigmatic and the Interpretive in Thomas Kuhn. *Clio* 17: 239–48.

Prigogine, I., and I. Stengers. 1984. *Order out of Chaos: Man's New Dialogue with Nature*. Toronto: Bantam Books.

Sacks, O. [1972] 1985. *Migraine: Understanding a Common Disorder*. Berkeley: Univ. of California Press.

———. 1974. *Awakenings*. Garden City, N.Y.: Doubleday.

———. 1984. *A Leg to Stand On*. New York: Summit.

———. 1985. *The Man Who Mistook His Wife for a Hat and Other Clinical Tales*. New York: Summit.

———. 1989. *Seeing Voices: A Journey into the World of the Deaf*. Berkeley: Univ. of California Press.

Serres, M. 1982a. *Hermes: Literature, Science, Philosophy*. Ed. J. V. Harari and D. F. Bell. Baltimore: Johns Hopkins Univ. Press.

———. 1982b. *The Parasite*. Trans. L. R. Schehr. Baltimore: Johns Hopkins Univ. Press.

History, Myth, and

Narrative

Stephen Jay Gould's Vision of History

LOUIS P. MASUR

With two out in the ninth inning of the 1956 World Series, umpire Babe Pinelli, on a count of one ball and two strikes against Dale Mitchell, called a strike on a pitch considered by most observers to be high and somewhat outside, thus giving Don Larsen the only perfect game in World Series history. Pinelli was reviled by many (mostly Dodgers fans) for making such a call, but he never harbored any doubts as to his judgment on that fall day. Following Pinelli's death several years ago, Stephen Jay Gould wrote a short piece in which he defends the umpire: "A batter may not take a close pitch with so much on the line. Context matters. Truth is a circumstance, not a spot" (1985, 227).

Gould's defense of Pinelli shocked some readers. How could a scientist appear not to believe in truth? Wasn't he advocating moral relativism? How do we navigate our way in the world if subjective perspective matters more than objective reality? Gould's critics overreacted to the piece and, in doing so, missed the larger point he was trying to make, a point that permeates all of his writings. The issue is not whether there is an external universe governed by unchanging natural laws; of course there is. No matter how many times Larsen threw that ball, it eventually would hit the ground. But facts and truths are not necessarily one and the same. To Gould, we are the products of a history that is continually being reimagined; history illustrates the contingent and contextual nature of the world in which we live (Gould 1989, 284–85).

Gould is one of a number of scientists who have continued in the tradition of popularizing discoveries in their field and writing for a wider public audience. Among recent examples, one thinks of the astronomer Carl Sagan, the physicist Stephen Hawking, and the physician Lewis Thomas. Others as well—Jerome Bruner, Freeman Dyson, Jeremy Bernstein, and Oliver Sacks—have labored to translate internalist scientific developments into public questions of both philosophical and practical significance. Whatever the problem—whether the origins of the universe, the consequence of nuclear explosion, or the battle with cancer—these writers have helped bridge the gap between laboratories, experimentation, and science, on the one side, and epistemology, narrative, and history, on the other.

No one has been more visible, successful, or influential in this endeavor than the evolutionary biologist and historian of science Stephen Jay Gould, professor of geology at Harvard University. His scientific papers have appeared in such professional journals as *Science*, *Paleobiology*, and the *American Journal of Science*. Over the past fifteen years he has written a monthly column for *Natural History*, and many of the essays that first appeared there are collected in four published volumes. He reviews for the *New York Review of Books* and writes more widely for such publications as *Discover*, the *New York Times*, and *Sports Illustrated*. He even contributed an introduction to the most recent collection of "Far Side" cartoons by Gary Larson. To call him prolific is an understatement. He has published so much, in so many different places, on so many topics, that a computer literature search using his name as author and employing data bases specific only to the humanities and social sciences generates a single-spaced printout that runs more than twenty-five pages.[1]

Gould's academic reputation was established through his fieldwork on the Bahamian land snail, *Cerion*, and his theoretical reconsiderations of the mechanisms of evolution. His popular reputation stems from an ability to make complicated scientific ideas accessible and relevant to a nonspecialized audience. Anyone who reads his essay on the evolution of Mickey Mouse's image (1980, 95–107), an essay that illustrates the scientific concept of neoteny (adult retention of childlike features) and illuminates shifts in American middle-class culture (Mickey goes from mischievous prankster to saintly child), cannot help but become hooked on Gould.[2]

Paleontologist, geologist, humanist, and critic—Gould is all of these. He is also—I am tempted to say primarily—a historian who writes elegant meditations on the history of evolutionary theory, the transmission of ideas, and the sociology of knowledge. At the core of Gould's work is a deeply historical perspective concerned with the fundamental issues of objectivity, interpretation, and causality. Discussion of these problems is hardly new, but it is noteworthy that while professional historians have been unable to capture a public audience for these theoretical questions, a scientist is playing a leading role in discussing historical methodology before a general readership.

An introduction to Gould's vision of history must begin with his perspective on what history is not. Foremost, history is not the objective reconstruction of the past, though many social scientists still hew to the belief that there is an empirical reality that can be objectively recovered. In this context, objectivity is usually juxtaposed against theory, the belief being that the objective researcher approaches his subject without theoretical assumptions. As far as Gould is concerned, nothing can be further from actual practice. In arguing for the deeply subjective and theoretical nature of the historical enterprise, Gould is not trying to perpetuate the traditional distinction between "soft" social sciences and "hard" sciences. Quite the contrary. Gould is calling on his own colleagues in geology and paleontology to "understand, appreciate, and use the principles of historical science" (1986b, 69). In part, this means recognizing that both disciplines are, to use the current parlance, socially constructed; both historians and scientists investigate problems and suggest solutions without shedding their political, social, and cultural skins.

Gould's primary concern is with fellow scientists who seem unable or unwilling to accept the conditional and contextual nature of their discipline. Yet, what he has to say applies with special force to humanists generally, many of whom continue to embrace the ideal of objectivity. It is important to note that Gould is not a complete heretic when it comes to science. He does believe in natural laws and scientific methodology; there are indisputable scientific laws governing the physical universe. But Gould is chiefly concerned with illuminating the interpenetrations between science and society, facts and fictions, nature and history.

Gould is not the first to emphasize that science is a social product; it has been nearly thirty years since Thomas Kuhn examined *The Structure of Scientific Revolutions* (1962) and argued that scientists work in communities where the questions they ask and the results they find are shaped by governing intellectual paradigms. One of Gould's most influential and controversial scientific papers opens with a quote from the Nobel laureate Peter Medawar that succinctly restates the same point: "innocent, unbiased observation is a myth" (quoted in Eldridge and Gould 1972, 84). It is astonishing how many scientists and social scientists refuse to acknowledge the validity of Medawar's apothegm: that one perceives what one expects to see; that the same sets of facts can be aligned toward dramatically different ends.[3]

The danger of the myth of objectivity is that it transforms theories into laws and elevates preconceived viewpoints into neutral truths. As Gould points out, Darwin himself fell victim to this process—both from others, who transmuted natural selection into a universal law governing the struggle for existence, and by his own hand when, in his autobiography, he claimed to have derived his theory from the arduous and systematic collection of facts though he knew full well that "all observation must be for or against some view if it is to be of any service" (Eldridge and Gould 1972, 85).

History is not neutral; it is an act of interpretation. And no interpretation troubles Gould more than that of gradual, deterministic progress. The idea of progress is one of the most powerful biases in Western thought. Needing to reduce the jangle of experience into a moral tale, we love nothing more than to speak of gradual improvement, of things getting better, of climbing to the top. Scientists who see themselves as objective experimenters searching for breakthroughs in knowledge (darkness into light is another favorite way of expressing the idea of progress) have been especially susceptible to this view of history. In evolutionary biology, the idea of progress has installed itself through the metaphor of the ladder (Gould 1989, 27–45). According to this way of conceptualizing evolution, humans occupy the highest rung on an evolutionary ladder that descends through billions of years. Evolution, in this view, becomes a story of progress, with some species higher on the ladder, others lower.

Because we have grown so accustomed to thinking and speaking in these terms, the perniciousness of the ladder metaphor is not readily apparent. Transferred from the evolutionary to the social realm, ladders become justifications for racism, inequality, and oppression. So ingrained was the belief that races had to be ordered in such a way that placed blacks at the bottom and whites at the top that scientists in the 1920s searched Asia for fossil remains that would provide evidence to contradict the indications that all human life originated in Africa. Selected members of the races and classes slotted at the bottom of the ladder had the promise of upward mobility held out to them should they follow social rules and conventions created by those at the top. In this way, the idea of gradual, upward progress contributed to sustaining the status quo. As Gould has pointed out, the understanding of evolutionary and historical changes in terms of gradualism and progress became commonplace in the nineteenth century not because these perspectives expressed objective, scientific data, but because as part of a Victorian ideology these watchwords became "liberalism's quintessential dogma against radical change" (Gould 1977, 207).

For Gould, history refutes this moralistic tale of gradual, uniform progress. History is complex, interactive, hierarchical, and, most of all, contingent. Simplicity, uniformity, and inevitability should have no place in scientific or historical visions; certainly the pretense that these are neutral conditions must be exposed. The more accurate metaphor for capturing the variability and diversity of evolution and history is the branching tree, the ramifying bush. With its multiplicities of divergences and interactions, its nonlinearity and dense complexity, the tree better represents both biological and cultural evolution. Family trees and social trees, not steps and ladders, encompass the genealogy and contingency of history (Gould 1977, 56–62; 1989, 45–52).

As a historical theorist, Gould does more than merely deride the illusion of objectivity and attack the ideology of progress; he attempts to reconceptualize approaches to evolution and change over time. Gould's theory of evolutionary change, first articulated with Niles Eldredge, offers an alternative perspective to an evolutionary pattern governed by creeping gradualism and preordained progress. Instead, Gould envisions the history of life as characterized by long periods of stasis that are suddenly and dramatically disrupted before there is a return to a new, stabilized order. In other words, punctuated equilibria. Sudden discontinuity and disruption, not gradual progress and slow ascent, characterize the tempo of historical change (Eldredge and Gould 1972, 1977; Gould 1982b).

Punctuated equilibrium is a theory of evolution devised to explain the histories of species. As a theorist of historical change, Gould owes an enormous debt to Georg Hegel and Karl Marx. Specifically, Hegel's dialectical laws ("interpenetrating opposites," "transformation of quantity to quality," and "negation of negation") speak to the ways in which revolutionary change grows out of apparent stability. Hegel's laws, as Gould has observed, "are explicitly punctuational." A dialectical approach, whether applied to evolutionary shifts that take place over hundreds of thousands of years or to social transformations that occur within a decade, provides a "holistic vision that views change as interaction among components of complete systems, and sees the components themselves not as a priori entities, but as both products of and inputs to the system" (1987d, 154).

Gould has identified some of the events to which the theory of punctuated equilibria can be applied. For example, recent evidence that attributes the disappearance of dinosaurs to a mass extinction caused by an asteroid striking the earth illustrates in dramatic fashion the principle of punctured stasis (1985, 230–44, 417–37). Applied to the less distant past, punctuated equilibria serves as an apt description of ideological and social revolutions. Consider the saying quoted by Gould that a soldier's life consists of long periods of boredom interrupted by short periods of terror (Eldredge and Gould 1977, 147), or contemplate the possibility of nuclear war, and one viscerally comprehends the meaning of punctuated equilibria.

It would be an error, however, to think of punctuated equilibrium only as a vision of dramatic transformation. Gould's historical perspective is as much about stability as change, as concerned with how orders are maintained as with how they are transformed; indeed, with the relationship between disruption and stability. The material and structural bases of any species (genes and environment) or society (economics and institutions) remain largely unaltered across long stretches of time. Understanding the mechanisms that perpetuate order ultimately helps us to understand how it is that orders are transformed. For students of history, this means paying close attention to the nature of power and authority in any social arrangement, to the myriad controls that keep the old order standing.

Gould's appreciation for deep structure does not in any way inhibit his passion for ideas. Among the numerous false dichotomies that permeate the Western mind, materialist versus idealist is one of the most persistent. But Gould refuses to succumb to the allure of dualisms (1989, 50–51). In his synthesis, ideas are matter, thoughts are actions. The bulk of Gould's work has been devoted to examining the formulation, expression, and transmission of ideas, from the heights of Sir Charles Lyell's uniformitarianism, Ernst Haeckel's ontogeny and phylogeny, and Charles Darwin's natural selection, to, in our own time, the depths of E. O. Wilson's sociobiology and Arthur Jensen's geneticism. As a historian of ideas, Gould understands that "the use of ideas, the systematic reconstruction of a world in their light, is the stuff of intellectual revolution" (1987d, 60).

Gould's dialectical approach requires that scientists and social scientists divorce themselves from simplistic, reductionist explanations. He has made this point repeatedly, but never more forcefully than in a paper, coauthored with Richard Lewontin, that attacks adaptationism (1979). The reigning paradigm in evolutionary thought for the past half century, adaptationism holds that most traits can best be understood as gradually, progressively attained ideal adaptations within species. The adaptationists' method is to atomize organisms into their constituent parts and then argue that each part fits optimally with the environment. Such a functionalist theory is not neutral; Gould and Lewontin have characterized it as "Panglossian," for it seeks to assure us that "we live in the best of all possible worlds." As a consequence, it reinforces the existing order with all of its divisions defended, even celebrated, as inevitable.

Proponents of a strictly adaptationist model, Gould and Lewontin point out, reason backward from parts to wholes, from uses to origins. They assume that all features (chins in humans or small arms in dinosaurs, for example) evolved for some functional reason and that this reason accounts for their existence. An example offered by Gould and Lewontin drawn from cultural rather than biological evolution is the argument that Aztec cannibalism can be explained as a creative adaptation to a shortage of protein. Such monocausal, reductionistic arguments are easily swallowed but not particularly satisfying. Organisms and culture are "not collections of discrete objects" but "integrated entities" (Gould and Lewontin 1979, 585). Adaptationism may explain some evolutionary features, but nonadaptationist processes, correspondences in structure and origin but not function (homology), and differential rates of growth (allometry) help illuminate many others. The point, for Gould and Lewontin, is that evolutionists must treat integrated wholes, not amputated parts; theorists of historical change must construct models that take into account the complexity of biological and cultural systems.

Gould is not so naïve or hypocritical as to pretend that his approach to evolutionary and historical change does not derive from his own worldview. Gould too writes from an ideological perspective, only in his case it is not so much traditional liberalism as democratic socialism. The connection between intellectual theories and political ideologies is a close one indeed. As Gould points out, it is not accidental that in the West Darwin is hailed as an advocate of gradualism, whereas in the East he is viewed as a theorist of revolution; Western scientists emphasize conflict as the central mechanism in natural selection, whereas their counterparts in the East stress the cooperative nature of species interactions. Gould's radicalism no more invalidates his approach to evolution than Darwin's conservatism undermines *Origin of Species* (1859). But it is likely that Gould's neo-Marxist perspective helps to explain part of the resistance to punctuated equilibria and nonadaptation among some evolutionary biologists (1988c, 12–21).

Gould is but one of a number of biologists who, in the last two decades, have brought a Marxist perspective to bear on their work. (Gould serves on the advisory board of the newly founded journal *Rethinking Marxism*.) In a recent volume entitled *The Dialectical Biologist* (1985), Richard Levins and Richard Lewontin elaborate on the dialectical approach. Like Gould, they argue that science is a "social process," that "the problematic of science—what questions are thought to be worth asking and what priority will be awarded them—is also strongly influenced by social and economic factors" (1985, 4). Social ideology and social relations cannot be divorced from scientific pursuits. Theorists, Levins and Lewontin argue, must try to break free from the stranglehold of Cartesian reductionism, of examining and understanding the world through its parts rather than as a whole. The dialectical approach focuses on the complex relationships and interpenetrations between parts and wholes, organisms and environments.

For these biologists, there is no dialectical method per se; to prescribe one would be to fall into the very trap they seek to avoid. Rather, their approach is to ask questions and conceptualize events in ways that acknowledge organic holism, historical fluidity, and social power. They are concerned with the relationships between the natural and the historical, the physical and the social. Dialectical laws are not "rules derived from nature" but intellectual principles and "terms of reference" (Levins and Lewontin 1985, 268). In this view, dialectics provides a way of approaching history as complex, interactive, and transformative.

Gould and Eldredge conclude their initial paper on punctuated equilibria with a plea that scientists become more sensitive to history. Scientists must recognize that ideas are culturally embedded; they must work in a present that is the product of a contingent past; they must appreciate the importance of narrative, of using words and images to tell stories that explicate change over time; they must craft multicausal explanations for historical events that are always complex, never simple. In axiomatic fashion, Gould has reiterated his message many times. The ideas that "history matters," that "everything, ultimately, may be a product of history," that "history . . . may be the ground for our search to understand cultural diversity and change" (1987d, 116) are at the core of Gould's perspective on human evolution.[4]

For Gould, interpretation of the past serves as a foundation for action in the present. In attacking the myth of objectivity and neutrality, in undercutting the distinction between thought and action, Gould and his fellow biologists have freed themselves to incorporate their moral values and social commitments into their work. Gould applies his evolutionary and historical understanding to social questions of enormous importance. The arguments he offers are applicable to some of the most intractable problems that face us today, and the positions he takes are consistent with those held by most activists on the left. The clearest example of this is Gould's concern with the problem of equality. Nothing roils him more than the use of evolutionary theory in defense of racial and social inequality.

In *The Mismeasure of Man* (1981), Gould scrutinizes the assumptions and methods of a number of scientists whose work promoted scientific racism and biological determinism over the past two centuries. The book is a deeply unsettling one for several reasons. In a society in which science was largely accepted by the public as objective and empirical, the individuals Gould discusses allowed their ideology to shape results. Thus Samuel George Morton's desire in the 1850s to create a ranking of races based on brain size led him *unconsciously* to tinker with the results of his measurements. Or consider Cyril Burt's more explicit manufacturing of data in the 1930s and 1940s to support his claim that intelligence is genetically determined.

As troubling as a history marked by subjectivity and fraud is, Gould does not stop there. He will not allow us the luxury of lapsing into a mythology that suggests we have progressed from those dark early days. The cultural context of science is as relevant today as it was in the nineteenth century. "If scientists can be honestly self-deluded," Gould writes, "then prior prejudice may be found anywhere, even in the basics of measuring bones and toting sums" (1981, 56).

The most striking chapter in *The Mismeasure of Man* (1981, 146–233) demonstrates how Alfred Binet, H. H. Goddard, and Lewis Terman, to one extent or another, developed a hereditarian theory of intelligence, indeed invented the concept of IQ, out of ideological motives and for social purposes. Numbers, Gould reminds us, "beguile and benumb" (1987d, 144). In an evanescent world they seem real, certain, unchanging. Yet faith in the objectivity of quantification is itself the invention of the nineteenth century, the first century to be consistently referred to by number. Numbers and their use are not benign. The obsession with calculating intelligence led directly to such practices as sterilization, institutionalization, and immigration restrictions for those suspected of being idiots or imbeciles (language employed as technical jargon earlier in this century). Long after these practices have ceased, intelligence testing persists, and the belief that intelligence can be quantified and that the number generated somehow stands for an objective, unbiased assessment continues to play a powerful role in society (Gould 1987a, 12–18).

Gould has not only attacked the idea of inherent inequalities, he has forcefully argued that the equality of all races is a biological fact, which is to say a "contingent fact of history" (1985, 185). By this he means that all races are members of the species *Homo sapiens,* that all descend from a common origin in Africa some tens or hundreds of thousands of years ago, and that the genetic differences between races are minimal. Inequality is a cultural, not a biological, phenomenon (1985, 185–98).

The contingency in Gould's motto refers to the possibility that other scenarios might have occurred. *Homo sapiens* evolved from one population of one line of *Australopithecus*. But what if another species of the genus *Homo* had survived until today and presented us with the "dilemma of a human species truly and markedly inferior in intelligence" (1985, 198)? What, Gould asks, if *Homo sapiens,* rather than being a geologically young species, had evolved over millions of years and exhibited truly deep differences within the species? History matters precisely because these things did not happen; the equality of races was not predetermined but the result of a historical process.

Despite the attempts of some to deny a common ancestral beginning in Africa, and despite an obsession among scientists until recently for distinguishing racial variations, the evolutionary history of humans is characterized by unity, connection, and equality. Gould recognizes the difficulty of convincing those who seek biological justifications for inequality that the argument cuts in precisely the opposite direction. How as well to make the leap from biology to ethics, from evolutionary patterns to cultural norms? For Gould, understanding begins with the recognition that races vary minimally, that individuals differ greatly, and that any individual deserves to be treated as "a full human being in all respects" (1985, 197). Human unity is literally true, not merely an "idle political slogan or tenet of mushy romanticism." "Our unities are genealogical," Gould concludes. "We are an object of history" (1988d, 21).

In his writings Gould has repeatedly shown the ways in which science has found regular employment as a justification for racism and inequality. So compelling is the associative power of the word *science* with truth and empiricism that in recent years opponents of the teaching of evolution in public schools have adopted the name *creation science* to describe their alternative vision. Gould has played a key role in opposing legislative acts that require the teaching of creationism alongside evolution. Having seen the last of the antievolution laws struck down by the Supreme Court in 1968, fundamentalists have adopted a pro-creation strategy. In the 1980s they were successful in states such as Arkansas and Louisiana in passing legislation requiring "equal time" in the classroom for creationism wherever evolution is taught. Only recently have the courts ruled that these acts were unconstitutional.

In the case of *McLean* v. *Arkansas Board of Education* (1982), Gould testified in federal court against an Arkansas act that required public schools to provide "balanced treatment to creation science and to evolution science." As a prominent opponent of the creationist legislative strategy, Gould was also closely associated with litigation over a similar statute adopted by the Louisiana legislature. In *Edwards* v. *Aguillard* (1987), the Supreme Court, by a vote of seven to two, found the Louisiana statute unconstitutional, thus ending a campaign that dates to the infamous Scopes Trial in 1925, designed to eliminate or mitigate the teaching of evolution in the schools.

If Gould has frequently attacked the idea of science as neutral, objective, and truthful, he has also defended the core principles of science, defined as "a system of explanation that relies upon invariant natural laws" (1982a, 10), against those who would eviscerate the entire discipline by trying to alchemize religious assertions and moral speculations into a science. As a student of history, Gould realized early that the fundamentalists' new legislative strategy against evolution had to be taken seriously and that their arguments, however transparent to the minds of secular humanists, had to be refuted systematically.

Gould has helped demonstrate that creationism is religion, not science. By emphasizing the origins of life, relying on supernatural intervention, and insisting on a literal interpretation of Genesis, creationism violates the cardinal tenets of a science: it ignores natural law, it is neither testable nor falsifiable, and it is dogmatic rather than explanatory. Furthermore, those beliefs of creationists that can be tested, such as the claim that the fossil record is a product of Noah's Flood or that the geological age of the earth is approximately six thousand years, are demonstrably false. Gould is no apologist for science, but after all, he comments, it "*has* taught us some things with confidence" (1987e, 34).[5]

Gould's elation over Judge William Overton's decision in *McLean* (Overton, who cited Gould's testimony in his decision, wrote "creation science is simply not science, . . . the purpose [of the act] is the advancement of religion in public schools") and the majority opinion in *Edwards* (Justice William Brennan ruled that the Louisiana Creationism Act "violates the Establishment Clause of the First Amendment because it seeks to employ the symbolism and financial support of government to achieve a religious purpose") has been tempered somewhat by one Supreme Court justice's misunderstanding of evolution.[6]

Justice Antonin Scalia, who along with Chief Justice William Rehnquist dissented in *Edwards*, fundamentally misconstrues the critical distinction between creation and evolution, a distinction that illuminates why one is not a science and the other is. Gould has shown that throughout Scalia's dissent the justice considered as part of evolutionary theory the study of life's origin. Creation and evolution are similar, Scalia suggests, because neither can account for the beginning of life. But evolutionary biologists attempt no such enterprise. Evolution merely declares that "all organisms are united by ties of genealogical descent" (Gould 1987f, 24); theorists of evolution say nothing about the origins of organisms from inorganic elements because such a question does not properly belong in the realm of science. More than a century after Darwin, Gould is saddened that he is still compelled to testify that evolution is not theory but fact. The theoretical component of evolution concerns only the mechanisms by which descent has occurred. By comparison, creationism is an untestable theory with no data to support its claims (Gould 1987b, 14–21).

That Justice Scalia either willfully or unknowingly tried to find a common ground between creationism and evolution is more than troubling to Gould. It shows that although the crusade to challenge the teaching of evolution has largely been halted, misunderstanding has not been eliminated. Creationists continue to press their assault on evolution, and only continued vigilance by teachers, scientists, and defenders of the separation of church and state, insists Gould, will lead to the ultimate extinction of creation science.

Unlike creationism, evolution teaches that there was nothing cosmically preordained or inevitable about the emergence of humans in history. On the bush of evolution, *Homo sapiens* is but a peripheral twig whose historical appearance is strikingly recent. Gould frequently quotes John McPhee's image for the shallowness of human roots. If all time on earth is represented by one's outstretched arms, "in a single stroke with a medium-grained nail file you could eradicate human history" (Gould 1987d, 3).

"One small twig" in evolutionary history is *Homo sapiens*. And yet, as a consequence of the development of consciousness—in itself "a quirky evolutionary accident" (Gould 1985, 431)—humans alone now hold the power of undoing billions of years of evolution and of possibly assuring their own extinction. This horrifying prospect has led Gould, along with dozens of other eminent scientists, to detail the biological consequences of nuclear war.

In 1983, Gould participated in a conference on the Long-Term Worldwide Biological Consequences of Nuclear War. (The following year he helped draft the Vatican statement on nuclear winter.) A committee of twenty scientists, including Gould, issued a report that discusses the likely consequences of a ten-thousand-megaton nuclear exchange; these include dramatically reduced temperatures, the disruption of photosynthesis, exposure to ionizing radiation, extensive chemical pollution, and the breakdown of temperate, tropical, and aquatic ecosystems. The scientists conclude that a nuclear war would leave no survivors in the Northern Hemisphere, and that the possibility of the extinction of *Homo sapiens* could not be ruled out (Ehrlich et al. 1983; Gould 1985, 435–37).

In an arena crowded with voices, Gould has added an evolutionary perspective to the horror of contemplated extinction. If such an event occurs,

> consciousness may not evolve again in any other lineage during the 5 billion years or so left to our earth before the sun explodes. Through no fault of our own, and by dint of no cosmic plan or conscious purpose, we have become, by the power of a glorious evolutionary accident called intelligence, the stewards of life's continuity on earth. . . . I cannot imagine anything more vulgar, more hateful, than the prospect that a tiny twig with one peculiar power might decimate a majestic and ancient tree, whose continuity stretches back to the dawn of earth's time, and whose trunk and branches house so many thousand prerequisites to the twig's existence. (1985, 431)

Gould's egalitarian and democratic impulses, his historical and scientific perspectives, have also led him to comment on political and social issues in which his expertise as a biologist matters far less than his worldview. As the grandson of Jewish immigrants and the son of a court stenographer who apparently was well versed in Marxist theory, Gould must have internalized a great deal about social class and political ideology, social discrimination and irrational persecution. One finds these concerns expressed in a variety of ways, from his self-consciousness over his position at an elite university to his participation in a panel on anticommunism in American intellectual life to his adding his name to a group of international figures who condemned United States policy toward Nicaragua and Soviet policy in Afghanistan as violations of "the democratic right of every nation to self-determination" (1986c, 44).

Gould is also a baseball fanatic. It should come as no surprise that, in his writings, Gould connects his interests. Baseball allows Gould to "apply" his evolutionary perspective while providing an arena that transcends the endless diversity of evolutionary biology for the heroic efforts and timeless rhythms of the playing field. In writing about baseball, Gould can discuss both the universal and the cultural, the history of life as well as the life of the individual.

Baseball is not just a game; it is the embodiment of history and time. It is both linear and cyclical and therefore fits nicely with Gould's recent work on time's arrow, time's cycle (1987c). The season begins in April and is measured through rapidly accumulating at-bats, innings, runs, games, wins, and losses until, in October, there are final standings for both individuals and teams. Through fall and winter the baseball field lies fallow (at least those that are still grass), until, with the return of spring, the cycle begins again. As Gould has expressed, "baseball fulfills both our needs for arrows (to forge time into stories) and cycles (to grant stability, predictability and place)" (1988b, A21).

In Gould's hands, our understanding of baseball is refined and enhanced through the prism of evolutionary theory. With its long stretches of scoreless equality ruptured by offensive outbursts, a baseball game serves as a compact illustration of punctuated equilibrium. Statistics reveal just how momentous these sudden runs-scoring rallies actually are. According to the *Bill James Baseball Abstract*, more than 60 percent of the teams that go on to win any given game score enough runs for victory in a single inning. In other words, more often than not, single-inning offensive displays determine the outcome of the game. Punctuated equilibrium helps us to comprehend the sudden, uneven, disruptive nature of historical change whether the occasion be asteroids hitting in the Cretaceous era or runs scoring in the top of the ninth.

A baseball game exemplifies punctuated equilibria, and the decline of .400 hitters illustrates another perspective drawn from evolutionary biology—variation. Ted Williams batted .406 in 1941, the year Gould was born, and since then no one has averaged over .400 for the season. By comparison, between 1901 and 1930 the league-leading average surpassed .400 nine times. The problem, simply put by Gould, is, What accounts for "the extinction of the .400 hitter"? (1985, 215).

Many of those who have considered the dilemma of the disappearance of the .400 hitter have relied on a perspective that can best be described as a theory of declension: once upon a time there were giants and heroes, and we are condemned forever to compare ourselves with them and come up wanting. For those inclined toward such a view, it becomes all the more disturbing and paradoxical to consider that by standards of training, diet, and performance against the clock, today's athletes are superior to those of half a century ago. To see the baseball players of the past as part of a golden age involves more than just nostalgia. As Gould puts it, it "carries moral implications linked metaphorically with junk foods, nuclear bombs, and eroding environments as signs of the current decline and impending fall of Western civilization" (1985, 216).

As a theorist of historical change, Gould rejects declension as "naïve and moral" just as he rejects progress as equally value-laden. Instead, he approaches the problem of the disappearance of the .400 hitter with a concept from evolutionary biology—the idea that "*trends in extremes* may result from systematic changes in *amounts of variation*" (1985, 219). As variation declines, extremes decrease; conversely, as variation expands, extremes increase.

If Gould is correct and a decline in variation helps explain a decline in extremes, then, applied to baseball, we should find a decline not only in the highest average but also in the lowest. Indeed, this is precisely the case. The point difference between highest and lowest batting averages compared with the general league average (which has remained surprisingly constant at about .260) has declined over time. What must be explained is not the decline of the .400 hitter, though posing the question in this way clearly tells us something about our own cultural prejudices, but the decline of variation in baseball.

Gould offers two interrelated explanations for the decline of variation in baseball. The first has to do with the "outer limits of human capacity" (Gould 1986a, 62): no trend goes on forever; there is always some outer limit to height, speed, life span. Gould would have us imagine a limiting wall for any performance. In baseball, the average performer has moved closer to the wall while the distance between ordinary and exceptional has decreased. Hitters who average .400 have disappeared not because players' skills have declined, but because they have improved. This leads to Gould's second explanation: "systems equilibriate as they improve" (1986a, 63).

As a consequence of repetition, standardization, and mounting precision, the game's rough edges have been smoothed. Better equipment, charting of pitches and hits, and positioning of fielders, to name but a few factors, have contributed to greater equilibrium in baseball and account for enough fewer hits for today's best batters to keep them from reaching .400. One extreme in a distribution of variation in batting average, .400 hitting, has disappeared as variation has declined, as the system of baseball has stabilized over time.

Baseball has become more scientific and, as with any science as far as Gould is concerned, it is subject not only to laws and statistics but to fictions and perceptions as well. Final scores and home runs may be counted as objective facts, but baseball also illustrates contingency and relativity, themes that permeate Gould's approach to the historical sciences. When Gould turns his attention from the universal and the historical to the individual, the connection between baseball and history becomes closer still (1985, 226–29).

If Gould has an evolutionary tree of heroes, the sturdiest branch belongs to Joe DiMaggio. In the summer of 1941 (again, the year of Gould's birth), DiMaggio did something never accomplished before or since in baseball: he hit in fifty-six consecutive games. Even those who know nothing of baseball somehow know of this record, and experts largely concur that DiMaggio's hitting streak is the single greatest individual achievement in baseball history, perhaps in all of sports, not to mention a defining episode in American cultural mythology.

For Gould, the majesty of what DiMaggio did between May 15 and July 16, 1941, can best be understood as an assault on nothing less than science itself. We live in a world governed by evolutionary laws, by inescapable probabilities, by statistical facts. As individuals, we live in history, and even as we reshape the past and present we succumb to a future that is both intentional design and unchanging structure. The wonder of DiMaggio's streak is that it shattered the laws of probability. He violated the predictions of random models that dictate with uncanny accuracy how long a streak will last given a specific individual's characteristic probability based on previous performance. DiMaggio smashed one history and created another. He showed that history and science are not objective, empirical, binding, or universal. History is a contest, a struggle, a constant effort on the part of the individual to understand and, if possible, to transform and transcend.

Gould's historical vision forces us to reconceptualize the world and our place in it. His perspective knocks us off the top of an imaginary ladder while revealing that we are deeply connected to all that surrounds us. His vision removes us from the center of a natural order thought to exist solely for our benefit without robbing us of the power to transform the world we live in. His theory even brings hope without defending the status quo, for it suggests that the longer things remain the same, the more dramatic will be the revolutionary change sure to come.

Gould is not alone in challenging the shibboleths of Western thought. Across numerous academic departments, scholars today are reconceptualizing their subjects and trying to build bridges between disciplines (Geertz

1983, 19–35). Social scientists and humanists are weaning themselves from the search for truth and are instead talking about narratives and stories; they are trying to move away from slender monographs to broad syntheses, from facts to theories, and from parts to wholes; not content merely to interpret, they are becoming activists and advocates seeking ways to apply their knowledge and understanding, searching for connections and audiences. It should come as no surprise that many of these writers are baseball fans as well.

As for DiMaggio, Gould summarizes the meaning of the Yankee Clipper's achievement this way:

> This history of a species, or any natural phenomenon that requires unbroken continuity in a world of trouble, works like a batting streak. All are games of a gambler playing with a limited stake against a house of infinite resources. The gambler must eventually go bust. His aim can only be to stick around as long as possible, to have some fun while he's at it, and, if he happens to be a moral agent as well, to worry about staying the course with honor. The best of us will try to live by a few simple rules: do justly, love mercy, walk humbly with thy God, and never draw to an inside straight.
>
> DiMaggio's hitting streak is the finest of legitimate legends because it embodies the essence of the battle that truly defines our lives. DiMaggio activated the greatest and most unattainable dream of all humanity, the hope and chimera of all sages and shamans: he cheated death, at least for a while. (1988e, 12)

NOTES

1. For an introduction to Gould, see Gleick (1983) and Tierney (1987). For a review of some of the major themes in Gould's work, see Sulloway (1987).

2. Gould's academic writings include his "Land Snail Communities" (1969), Schindel and Gould (1977), and Gould and Calloway (1980). Gould wrote about *Cerion* for his one hundredth essay in *Natural History;* see *The Flamingo's Smile* (1985, 167–84).

3. For a history of the ways in which historians have approached the problem of objectivity, see Novick (1988).

4. For a penetrating discussion of the difference between science and history, see Berlin (1979, 103–42). Gould dedicated *An Urchin in the Storm* to Berlin and Medawar.

5. See also Gould (1988a; 1983, 253–90).

6. Overton's decision in *McLean* was reprinted in *Science* 215: 934–43. The *Edwards* decision can be located at 107 S.Ct. 2573 (1987). For overviews, see Lafollete (1983) and Larson (1985).

REFERENCES

Berlin, I. 1979. *Concepts and Categories: Philosophical Essays.* New York: Viking Press.

Ehrlich, P., et al. 1983. Long-Term Biological Consequences of Nuclear War. *Science* 222: 1293–1300.

Eldredge, N., and S. J. Gould. 1972. Punctuated Equilibria: An Alternative to Phyletic Gradualism. In *Models in Paleobiology.* Ed. T. J. M. Schopf. San Francisco: Freeman, Cooper.

———. 1977. Punctuated Equilibria: The Tempo and Mode of Evolution Reconsidered. *Paleobiology* 3: 115–51.

Geertz, C. 1983. *Local Knowledge: Further Essays in Interpretive Anthropology.* New York: Basic Books.

Gleick, J. 1983. Stephen Jay Gould: Breaking Tradition with Darwin. *New York Times Magazine*, 20 November.

Gould, S. J. 1969. Land Snail Communities and Pleistocene Climates in Bermuda: A Multivariate Analysis of Microgastropod Diversity. *Proceedings of the North American Paleontological Convention.* Part E: 486–521.

———. 1977. *Ever since Darwin: Reflections in Natural History.* New York: W. W. Norton.

———. 1980. *The Panda's Thumb: More Reflections in Natural History.* New York: W. W. Norton.

———. 1981. *The Mismeasure of Man.* New York: W. W. Norton.

———. 1982a. Genesis vs. Geology. *Atlantic Monthly*, September.

———. 1982b. The Meaning of Punctuated Equilibrium and Its Role in Validating a Hierarchical Approach to Macroevolution. In *Perspectives on Evolution.* Ed. E. R. Milkman. Sunderland, Mass.: Sinauer Associates.

———. 1983. *Hen's Teeth and Horse's Toes: Further Reflections on Natural History.* New York: W. W. Norton.

———. 1985. *The Flamingo's Smile: Reflections on Natural History.* New York: W. W. Norton.

———. 1986a. Entropic Homogeneity Isn't Why No One Hits .400 Any More. *Discover*, August.

———. 1986b. Evolution and the Triumph of Homology, or Why History Matters. *American Scientist* 74: 60–69.

———. 1986c. Letters. *New York Review of Books*, 13 February.

———. 1987a. Bushes All the Way Down. *Natural History*, June.

———. 1987b. Justice Scalia's Misunderstanding. *Natural History*, November.

———. 1987c. *Time's Arrow, Time's Cycle: Myth and Metaphor in the Discovery of Geological Time*. Cambridge: Harvard Univ. Press.

———. 1987d. *An Urchin in the Storm: Essays about Books and Ideas*. New York: W. W. Norton.

———. 1987e. The Verdict on Creationism. *New York Times Magazine*, 19 July.

———. 1987f. William Jennings Bryan's Last Campaign. *Natural History*, November.

———. 1988a. Genesis and Geology. *Natural History*, September.

———. 1988b. Innings. *New York Times*, 4 April.

———. 1988c. Kropotkin Was No Crackpot. *Natural History*, July.

———. 1988d. A Novel Notion of Neanderthal. *Natural History*, June.

———. 1988e. The Streak of Streaks. *New York Review of Books*, 18 August.

———. 1989. *Wonderful Life: The Burgess Shale and the Nature of History*. New York: W. W. Norton.

Gould, S. J., and C. B. Calloway. 1980. Clams and Brachiopods—Ships That Pass in the Night. *Paleobiology* 6: 383–96.

Gould, S. J., and R. Lewontin. 1979. The Spandrels of San Marco and the Panglossian Paradigm: A Critique of the Adaptationist Programme. *Proceedings of the Royal Society of London* 205: 581–98.

Kuhn, T. S. 1962. *The Structure of Scientific Revolutions*. Chicago: Univ. of Chicago Press.

Lafollete, M. C. 1983. *Creationism, Science, and the Law: The Arkansas Case*. Cambridge: MIT Press.

Larson, E. J. 1985. *Trial and Error: The American Controversy over Creation and Evolution*. New York: Oxford Univ. Press.

Levins, R., and R. Lewontin. 1985. *The Dialectical Biologist*. Cambridge: Harvard Univ. Press.

Novick, P. 1988. *That Noble Dream: The Objectivity Question and the American Historical Profession*. Cambridge: Cambridge Univ. Press.

Schindel, D. E., and S. J. Gould. 1977. Biological Interaction Between Fossil Species: Character Displacement in Bermudian Land Snails. *Paleobiology* 3: 259–69.

Sulloway, F. J. 1987. The Metaphor and the Rock. *New York Review of Books*, 26 May.

Tierney, J. 1987. Stephen Jay Gould: The *Rolling Stone* Interview. *Rolling Stone*, 15 January.

Chaos out of Order: The Writerly Discourse

of Semipopular Scientific Texts

ROBERT T. KELLEY

When Benoit Mandelbrot titled one of the sections of *Fractals: Form, Chance, and Dimension* (1977) "This Essay Mixes Styles; Is Semipopular *and* Scholarly" he was doing more than attempting to describe his unique practice. This section clearly demonstrates the fractal mathematician's uneasiness concerning his place in the community of scientific discourse as he asserts that his text "will show tolerance regarding the numerous compromises that are unavoidable whenever one mixes styles" (1977, 22). Mindful that scientists might not take seriously his innovative work combining mathematics and computer graphics, which has subsequently become so familiar, Mandelbrot assures us that *Fractals* is a serious "work of erudition" (1977, 22).

He explains that "this work is in part expository. But this is not its main purpose. . . . this Essay has taken on some of the appearances of a work of popularization. But this is also not its main purpose" (1977, 22). As Mandelbrot's apprehension about his mixed discourse shows, the process of presenting scientific ideas is not always as straightforward as we might expect; some nonscientific elements, particularly narrative (what Mandelbrot thinks of as the "popular"), must enter into the discourse of works on the cutting edge of a new field. The popular, as Mandelbrot self-consciously conceives of it, is linked to the expository; it is descriptive rather than prescriptive.

In contrast to Mandelbrot's anxiety, Ilya Prigogine and Isabelle Stengers comfortably seek to bridge the perceived gap between scientific and expository discourses in *Order out of Chaos: Man's New Dialogue with Nature* (1984). While describing the history of and present advances in the study of chemical and physical systems that, when operating in a chaotic fashion, suddenly form regions of order (what they call "dissipative structures"), they seek to make the important general implications of their work accessible to a broader audience. The text sets up, through background philosophy and history, a framework for appreciating the findings the authors report in the final, more technical, section of the book. As the title of the earlier French edition of the text, *La Nouvelle Alliance* (1979), indicates, Prigogine and Stengers seek to describe how recent developments in the study of chaotic systems have suggested new ties between the sciences and the humanities. *Order out of Chaos* is not only a text that popularizes a scientific concept, it is also one that actively seeks to bridge the gap between the popular and the scientific.

An analysis of these texts must therefore take the conventions of both scientific and narrative discourse into account, because both Mandelbrot and Prigogine and Stengers actively work in both forms. Mandelbrot's discomfort in characterizing the form of his own work and Prigogine and Stengers's obvious preoccupation with the character of theirs foreground a significant concern in dealing with any text about science: How do we analyze a text that hovers between disciplines—is both semipopular and scientific (or historical or even literary)? These texts are not strictly popular, since they attempt to present real scientific findings in this format, yet they are certainly not traditional scientific textbooks.[1] Mandelbrot's description of his text as semipopular and scholarly certainly seems to describe the intended reception of the text, but it suggests nothing of its structure or strategies.

"Semipopular" suggests a form of analysis that would consider such texts as narrative in much the same spirit that Hayden White considers historical narratives to be "verbal fictions, the contents of which are as much *invented* as *found* and the forms of which have more in common with their counterparts in literature than they have with those in the sciences" (1978, 82). We can make a parallel assertion for traditional scientific texts: such texts are verbal fictions, the contents of which are as much invented as found and the forms of which are isomorphic with those of narrative texts. While the traditional scientific paper seems devoid of the rhetorical play and obvious fictionality of the novel, the semipopular science text lies somewhere in between traditional scientific and novelistic texts. In fact, the semipopular text's combination of discourses reveals the narrative that is science, the way in which science is constructed. But it also points toward the slippery politics of writing science. The rhetorical strategy of a traditional scientific text is to make an assertion and then to indicate that this assertion follows obviously from the evidence given. The traditional text elides the failed experiments and deadend hypotheses. It also puts up a unified front, implying that only one interpretation of the results is possible, that only one is called for. The strategy of the traditional text is to close down alternative possibilities and, through rhetoric, to indicate that the one finally chosen is the obvious and logical one. By contrast, the semipopular scientific text allows narrative into the arena, yet, in doing so, shows all the more how acknowledging that the story of science is a "verbal fiction" does not change the fact that it is a discourse that inevitably drives toward unity and closure but fails to attain it.

In order to see scientific discourse's inexorable move toward—but failure to reach—closure, it is useful to examine its rhetorical strategies. Alan Gross applies a modified concept of classical rhetoric to our conception of scientific knowledge in *The Rhetoric of Science* (1990), arguing that "scientific knowledge is not special, but social; the result not of revelation, but of persuasion" (20). In a statement echoing Hayden White, he contends that "scientific discovery is properly described as invention," where the term *invention* "captures the historically contingent and radically uncertain character of all scientific claims, even the most successful" (7). The way in which these discoveries are presented—the rhetorical device that makes them believable—Gross calls overdescription, which he defines as "the characterization of sense objects in detail far beyond a reader's ordinary expectations" (42–43). Along with other rhetorical devices, overdescription helps to create a "referential presence" (44), a sense that scientific language refers unproblematically to the real world. The rhetorical view of science does not privilege scientific knowledge, but it does acknowledge that the gathering of knowledge in Western philosophy is based on rhetorical strategies, and therefore all of the knowledge we take to be true is manufactured via rhetoric. "The rhetorical view of science does not deny the 'brute facts of nature'; it merely affirms that these 'facts,' whatever they are, are not science itself, knowledge itself" (4). Instead, knowledge is the answer to questions dealing with the interpretations of these facts. The processes "by which problems are chosen and results interpreted, are essentially rhetorical: only through persuasion are importance and meaning established" (4). The rhetorical view therefore studies "the world as meant by science" (4). The narratives of science, especially popular science, are stories whose persuasive intent is to make themselves believable, to pass off whatever degree of fictionality they have as truth. The author's aim—to make the narrative believable through overdescription—is thus similar to the aim of the author of the classical realist novel, who gives enough seemingly meaningless detail to make the text appear realistic. According to Roland Barthes, the "effect of the real" is produced through the addition of many "futile" details, which increase the "cost of narrative information" (1986, 141). These significantly insignificant details make the text seem like real life, in which not everything one encounters can be meaningful. The parallels indicated between a rhetorical view of science and Barthes's narrative theory therefore prompt us to consider scientific texts in terms of narrative.

Colin MacCabe's definition of the classical realist text "as one in which there is a hierarchy amongst the discourses which compose the text" where "this hierarchy is defined in terms of an empirical notion of truth" (1985, 34) suggests how we can begin to compare scientific and literary texts. This hierarchy is quite similar to the structure within the scientific text, one which suggests either that experimental evidence leads "directly from sensory experience to reliable knowledge about the world" or that theoretical formulations truly represent reality (Gross 1990, 91, 94). This kind of hierarchy is best captured in Roland Barthes's concept of the readerly text as he develops it in *S/Z* (1974). Barthes opposes the teleological *readerly* text to the *writerly* text, which demands intervention by the reader. The writerly text is characterized by plurality and polysemy; it forces the reader to make meaning. The readerly text, on the other hand, attempts to close meaning, to force the reader into a single interpretation. This perpetuates what Barthes has elsewhere called "doxa" (1975, 28), ideological statements that pass themselves off as nature, as already given. The reader of the readerly text is a passive reader, a reader "left with no more than the poor freedom to either accept or reject the text" (1974, 4). According to Barthes's scheme, these two poles of reading possibility are actually modes of meaning production—the writerly mode is one in which the reader interactively assists in the production of meaning; the readerly mode is one in which the reader merely accepts the previously produced meaning.

The text Barthes analyzes in *S/Z*—Honoré de Balzac's novella *Sarrasine*—is essentially a mystery story, one of the premier forms of teleological narrative. Barthes describes the function governing the building up of suspense or the unraveling of mystery in terms of the "*hermeneutic code*, all units whose function it is to articulate in various ways a question, its response, and the variety of chance events which can either formulate the question or delay its answer; or even constitute an enigma and lead to its solution" (Barthes 1974, 17). *Sarrasine* relates the story of the slow revelation of the sexual identity of a castrato to a young man (Sarrasine) who had fallen in love with the castrato dressed as a woman. The story possesses many twists and turns and is structured like many other suspense stories. It is obvious from both the internal story and the frame story that a predetermined ending has been established and that the getting to the end, the suspense that such a journey entails, provides the pleasure in reading. The reader, unaware of the outcome but delighted by the narrative which gets her there, feels a sense of completeness in the way the story closes. The story and its reading both invoke a kind of contract: in *Sarrasine* a man desiring sexual favors tells the tale of young Sarrasine, and in the reading process the reader desires pleasure and seeks it in the enjoyment of reading a story. The scientific text involves a similar contract. The reader does not consider the fact that the logical and certain findings of a paper could be disproved the following month; this "suspension of disbelief" is expected of the reader (Gross 1990, 95). The structure of the experimental report "recapitulates a movement from the contingency of laboratory events to the necessity of natural processes; in other words, the arrangement of the experimental report reenacts the process of induction" (Gross 1990, 86). As with the mystery story, we can posit a pleasure derived by the reader of the scientific paper as his or her beliefs about the outcome and relevance of a particular experiment are confirmed.[2] It is just this kind of comforting teleology, the neatly tied ending, that *Sarrasine* and the scientific essay both possess, and which Barthes subsequently deconstructs.

As a very carefully plotted story, *Sarrasine* is focused on the surprise ending. The tale is determinate—at no point in the story would it occur to the reader that either the narrator or Balzac himself is ignorant of its conclusion. Barthes, on the other hand, comments that this carefully constructed tale is, in fact, self-contradictory in places, that the text opens up holes in its careful plot. As a tale of mistaken identity, it questions the whole concept of identity: How do we know who is who, or, for that matter, what is what? While this question may seem glib, Barthes is able to demonstrate through careful analysis that the identification of characters is a task the reader must perform; meaning must be made by the reader, and the reader therefore assists in the construction of textual meaning. If, in the readerly text, we extract meaning according to a subtle but powerful agenda that the author has provided for us, Barthes's analysis of *Sarrasine* demonstrates that in the very act of reading, we are making meaning as we go; we are taking various cues from the text, not all of which the author necessarily intended.

The scientific text, of course, is not open to interpretation in the same way as is the literary text. The connotative richness of literary language is not supposed to exist in the scientific paper, nor are rhetorical figures acceptable in scientific scholarship except in the early stages of a theory—and even then only as heuristic devices (Gross 1990, 80). As readers, we are supposed to follow the author's stated agenda and not deviate from it in the slightest detail. Yet, reference in scientific discourse operates in essentially the same way as it does in the discourse of the classical realist text. While scientific prose appears to refer unproblematically to the causal structure of the world, scientific knowledge is as much "a matter of persuasion and consensus" as any other kind of knowledge (Gross 1990, 194). Even in agreeing with the premises, method, and conclusions of an argument, the unquestioning reader still participates in the process of creating meaning. Any statement about the method or conclusions of a paper is therefore fundamentally an interpretive act.

Thus, the acceptance or rejection of a text depends on the shared assumptions of reader and author. A literary author takes advantage of certain reading and cultural codes that he or she thinks we, as readers, will share. Reading and correctly interpreting these codes, we garner the intended meaning of the text. The author assumes, therefore, that these codes can be placed clearly in language and interpreted correctly and consistently by readers. Yet, while these codes are present in the text, since the author is immersed in the very culture whose codes he or she is supposed to be able to manipulate, the author's conception of the text may be subsequent to the codes; that is, their influence may precede the text (Barthes 1974, 140). If this is true, then can we as readers ever really tell who is speaking—the determined narrator or the culturally bound author? It is a case of asking, Which came first, the language or the author? If the author passes on codes about which she or he is not aware, then she or he cannot produce a fully closed text, a readerly text; some ambiguity and polysemy will still exist in the text. A reader, who may or may not share the author's culture, who subsequently sees the manipulated codes and interprets them independently, acts in a writerly way, producing unique meaning. The readerly text, then, has features that undermine its closed identity: no matter how hard an author tries to fix the meaning of a text, to make its conventions appear natural, its polyvalent qualities remain present; and no matter how careful an author is, he or she cannot control all elements of the text.

Barthes explores these limitations and possibilities as part of fictional narratives, not as elements of descriptive or explanatory texts, yet we can see obvious parallels to scientific texts. In referring to scientific textbooks, Thomas Kuhn asserts that "they have to be rewritten in the aftermath of each scientific revolution, and, once rewritten, they inevitably disguise not only the role but the very existence of the revolutions that produced them" (1970, 137).[3] The scientific community looks to textbooks in the same way that readers (according to Barthes) look at classical readerly narrative; expecting a single interpretation, the reader is convinced that the unified account that is read is innocently representative and "natural." Kuhn, like Barthes, points to ways in which writers participate in a conscious effort to cover up inconsistencies, to perpetuate doxa. This convergence should indicate to us that scientific texts, though different in style, are perhaps prone to the same kinds of forced teleology as are narrative texts. Popular scientific texts certainly have more in common with both literary and historical narratives than textbooks do, but we must consider to what degree these mathematically oriented, empirical texts are also writerly. How much does their language affect scientific thought—how implicated are scientists in the conventions of either natural or formal language?

Work in the philosophy of science and in studies in the gestalt nature of observation suggests that even in simple cases, what a scientist expects to see determines what is actually seen, that any observation or statement is theory-laden.[4] Similarly, the poststructuralist conviction that we are determined by or immersed in language is in various ways applicable to scientific texts. The view that even mathematics, that seemingly unassailable formal language, is subject to point of view or may be created and not found must be examined. Bruce Gregory neatly summarizes this view in *Inventing Reality* (1988) when he comments that "ontology recapitulates taxonomy—the way we divide the world in language tells us how we think the world is 'really' put together" (174). He demonstrates this concept using a discussion of Euclid's fifth axiom and shows that the interpretation of events through the language of Euclid's axioms severely limited the possibilities for geometry. The analysis of alternative possibilities produced Riemannian and other non-Euclidean geometries (which, ironically, better describe portions of the physical universe). Gregory goes on to conclude that

> mathematicians think of a mathematical system as a formal language that can be interpreted by providing a model of the system. For a physicist a mathematical system is a model of the physical world. For each of them an abstract language can be mapped onto features of the

world in a variety of ways. But there seems to be no unique way of combining a mathematical language and a part of the world. (177–78)

In other words, while the fact that the physical world is a model for mathematics (it serves as an example that verifies the theoretical formulations) demonstrates that mathematics does work for isolated physical events, there is no absolute relationship between formal language and physical reality.

Gregory points out an additional problem for twentieth-century physics. Given that the observer and the observed particles are inextricably tied together in postquantum physics, the myth of the detached observer and of an independent reality had to be overturned (181).[5] Regardless of the existence of events and objects at the supraatomic level, quantum physics and the recent developments in the chaos sciences demonstrate that the observations we record and explain are determined by language. The language of science and mathematics tends toward the determinate, yet its power to suggest has given us glimpses of a world that is indeterminate. Physics is a language, a precise and very fruitful one, but only one possible language for which the physical world is a model. The relationship between the mathematical-linguistic representation and reality is a fortunate or fruitful but not necessary correspondence. Since mathematics is an incomplete model, and since we are confident that narrative language no more points to true and unique objects than mathematics does, we can surmise that no language in the physical sciences indicates unique real objects, nor does a language ever fully describe a state of reality. Because we realize that the choice of language or of representative method largely determines what we will see, we know that language is implicated in our formulations of how the world is constructed. This realization entails more than indeterminacy: it is an inextricable linking of representation and the view of reality that both precedes and follows from it.[6]

Scientific texts, therefore, seem susceptible to the same kinds of analysis that Roland Barthes performs on *Sarrasine*. In short, despite its very different appearance and seemingly different assumptions, scientific discourse shares many basic traits with realist narrative. We can summarize these comparisons by positing the scientific text to be similar to the readerly text in four ways. First, it attempts to close meaning through the use of authorized denotative meaning. Second, it attempts to make the conventions of representation cover a particular interpretation, to make incomplete representation look like nature, like doxa. Third, even the most seemingly closed text possesses or, actually, incites the reader to produce, additional meanings, often in conflict with the authorized one. And finally, language choice precedes the production of the text, shaping it outside the author's active control rather than serving as a transparent tool for the author to use to directly convey meaning.

A rhetorical view of science addresses and clarifies each of these problems in relation to scientific texts because

> the objectivity of scientific prose is a carefully crafted rhetorical invention, a nonrational appeal to the authority of reason; scientific reports are the product of verbal choices designed to capitalize on the attractiveness of an enterprise that embodies a convenient myth, a myth in which, apparently, reason has subjugated the passions. But the disciplined denial of emotion in science is only a tribute to our passionate investment in its methods and goals. (Gross 1990, 15)

Scientific prose attempts both to close meaning completely and to pass that meaning off as unproblematically true or real through conventions of language. That which we call science is that which closes off meaning, that which appears neutral. In what is almost a tautology, then, science is the best prose for describing reality because we have decided to look at reality scientifically—assigning its representation an ethical neutrality which is then echoed by scientific language. Within closed scientific texts, however, is the potential for more than one interpretation, for additional meaning outside the authorized one. This fact, especially in the scientific text, indicates the importance of discovering the text's embeddedness in language. The rhetorical view of science exposes the manner in which scientific knowledge is manufactured through persuasion, through the choice of which details to look at and how to arrange them.

The writerly mode can be difficult to discern in scientific discourse, but it becomes apparent when the production of a scientific narrative is examined. Alan Gross's consideration of peer review processes show the ways in which the history of controversy surrounding a text becomes buried in the final scientific report. He claims that as a result of the process of peer review, in which an author must accept the changes demanded by a review committee made up of his or her peers,

> there is a cultivated systematic neglect of the relationship between the claims in these reports and the process by which their truthfulness is initially certified: the move to publication systematically distorts the

wholly argumentative grounding of the knowledge that the peer review certifies. In the public domain, all that is visible is the scientific
report, another step in the steady march toward certain knowledge.
(1990, 140)

This analysis can be applied to the historical evolution of book-length
scientific texts. The nature of the development of readerly discourse in a
scientific text is particularly obvious in a study of the history of Benoit Mandelbrot's revisions of his fractals texts, especially the two English editions:
Fractals: Form, Chance, and Dimension (1977) and *The Fractal Geometry
of Nature* (1983). As one of the earliest books on chaos studies, *Fractals*
holds a unique place in scientific discourse. Kuhn contends that one of the
unmistakable markers of a new discipline is a text that is intelligible to an
educated general audience; after its appearance, most important work in
the field is done in very specific and technical articles in learned journals.
The status of a textbook, then, is the benchmark of how far a particular
field has come. Its standard structure, with the historical reduction at the
beginning, serves to solidify the basis for additional research in the field.
All subsequent creative research uses the principles outlined in this beginning text and addresses specific and esoteric topics (Kuhn 1970, 19–20).
This standard textbook, then, cannot be ambivalent about the principles of
the new field; it must attempt the kind of closure indicative of the readerly
or classical text in order to perpetuate the ideas of the new paradigm.

   Mandelbrot opens *Fractals* not with a history but rather with an apology,
claiming that "this work is referred to throughout as a scientific Essay,
and it conforms indeed strictly (other than by its length) to an old dictionary's definition as a 'composition dealing with a subject from a personal
point of view and without attempting completeness'" (1977, 2). Obviously, Mandelbrot sees the difference between his own work and the work
of colleagues in more stable disciplines. He elaborates at length on this
difference, asserting that his work

   should help unify what has been until now a collection of mathematical odds and ends, many of them classical but somewhat obscure, and
   make them known to nonmathematicians and mathematicians alike.
   Also, the mathematical questions it raises are compelling on their own
   merits. However, this Essay is neither a treatise nor a textbook in
   mathematics.
      It is a collection of theories in diverse branches of natural science,
   that should be judged by the prospective users on the basis of their

powers of organization and explanation (and their esthetic quality) rather than by their attractiveness as examples of a mathematical structure. . . . this Essay is all preface from beginning to end. (1977, 2)

Mandelbrot apparently feels the pressure of what Kuhn describes as the transition into a scientific paradigm and responds by setting up a genealogy, creating a history. In fact, in the very next paragraph Mandelbrot begins the standard historical account, starting with Georg Cantor and continuing through now familiar names such as Helge von Koch and Waclaw Sierpinski. Mandelbrot rightly claims (in 1977) that these are names "not ordinarily encountered with any frequency in the empirical study of Nature" (1977, 2); though, as N. Katherine Hayles notes, Mandelbrot's progression in the subsequent revision of *Fractals* into *Fractal Geometry* is toward covering up this disparity and creating a more unified history which quite conspicuously includes himself (1990, 168).

Fractals, then, is an excellent example of an early attempt to begin to close a field, to bring previously unconnected information into one place and unify it into a single, readerly account. Mandelbrot wastes no time trying to conventionalize the study of fractals, attempting to make their application seem natural: "under the formal shell that has insulated [this collection of mathematical sets], its basic ideas are extremely simple and intuitive" (1977, 14). Like so many narratives whose conventionality becomes obvious, Mandelbrot comments that, with sciences, "it should be a matter of regret that the least exact among them, sciences whose very principles are the least certain, tend to be the most concerned with rigor, generality, and axiomatics. Thus I am delighted to feel I may have identified fresh examples in which, in the classical fashion, form seems to be intimately related to substance" (1977, 14). Here, and in many of the examples that follow, Mandelbrot claims to describe already existing mathematical entities whose true import only he has brought to light—which he has not created but rather discovered. This move is neatly covered over in more paradigmatically advanced texts whose history is already fully conventionalized; the way these sets are forced together into a science is arbitrary, yet it is pictured by Mandelbrot as natural. The change in titles alone in the revision points toward this shift—from the analytical title *Fractals: Form, Chance, and Dimension* to an objectively descriptive one, *The Fractal Geometry of Nature*. Also obvious in these passages is the authorial/authoritative voice that characterizes closed or readerly discourse; Mandelbrot's solicitous style—as he tells the reader what difficult parts can be skipped—is actually just as dominating a style as the realist narrator's. We only know what Mandelbrot tells us, and what he tells us is all there is to know. This aspect is especially obvious in the early sections defining the word *fractal,* when Mandelbrot does not hide his authority: "this etymology can be asserted with full authority because I am responsible for coining the term to denote a collection of concepts and techniques that seems finally to acquire a clear-cut identity" (1977, 4).

Mandelbrot claims that he is writing *Fractals* to introduce these concepts to scientists and other interested parties so that they can use them, yet this move toward a unified account, a single definition, already closes off some meaning. Any scientist or writer who discusses fractals or mathematically chaotic systems after Mandelbrot must both doff to him and defer to his definitions. As James Gleick indicates, Mandelbrot insists upon his status as founder of the field: "At the height of his success, he was reviled by some colleagues, who thought he was unnaturally obsessed with his place in history. They said he hectored them about giving due credit. . . . Sometimes when articles appeared using ideas from fractal geometry he would call or write the authors to complain that no reference was made to him or his book" (1987, 111). Mandelbrot's own feelings on this subject are clear: "No one disputes me full credit for this discovery. After all, I was in the wilderness a long, long time while people laughed at my ideas. . . . That the theory is mine is never argued. However, scientists in various disciplines expand on my work and my share of the whole will inevitably decrease. This is as it should be" (quoted in Prince 1984, 52). This hierarchy of theories, based on a system not unlike a kind of patrilineage, already limits any work in this field after Mandelbrot's.

In fact, as James Gleick's Pulitzer Prize–winning *Chaos: Making a New Science* (1987) would indicate, a history of chaos without Mandelbrot is unthinkable. Mandelbrot's audacity shows that what is taken as the truth in a field is as much a construction of who does the telling as what is told. For example, Gleick's "historical" narrative ignores almost completely Ilya Prigogine's Nobel Prize–winning work on chaos. As David Porush points out, however, Prigogine's project is different in some ways from the one in which Gleick sees Mandelbrot engaged, or in which Mandelbrot sees himself engaged. Whereas Mandelbrot is clearly involved in a kind of Kuhnian process of making his text the textbook of a paradigm, Prigogine is developing "a richer view of science as a cultural phenomenon" (Porush 1990, 437). Prigogine is more interested in the broad implications of his theory, the ways in which it touches on other disciplines and points of view. Mandelbrot sees himself "making a new science," as is apparent from his involvement in numerous priority disputes and his insistence that his book "has become something of a cultural phenomenon" (Prince 1984, 52). Prigogine, on the other hand, has an even larger project in mind; he sees his work as embedded in a history of questioning, a philosophical framework (Porush 1990). He does not wish to close meaning in the same way that Mandelbrot does, nor does he attempt to hide his own indebtedness to previous work, as Mandelbrot does.

As Barthes's work indicates, however, the harder one attempts to close meaning, the more difficult it becomes to hide the potential polysemy of the text. Even the style of Mandelbrot's book indicates that the subject of fractals is beyond the unifying control that Mandelbrot wields in one way but is unable to retain in another. The readerly text controls its topic, yet, as with studies in quantum mechanics, texts on the chaos sciences are unable to be definitive in the way in which the authors might intend. For example, as in most of the books available on chaos, the centerpieces of Mandelbrot's book are the stunning computer-generated graphs of the chaos functions.[7] The graphs of these functions have become a veritable trademark for chaos studies—most major texts on chaos have one on the cover and a number as plates. These graphs are not informative in the same way graphs of physical variables might be; rather, their comprehension requires a more general, almost aesthetic, appreciation. Mandelbrot's knowledge of the way in which the pictures and graphs function differently in his work than in other scientific texts could not prevent their connotative associations from exceeding his control. In spite of his debates both with artists using his work (in his arguments with Lucasfilm over how to best generate fractal planetscapes) and with colleagues not using his work (an ongoing problem of failure to properly acknowledge him), Mandelbrot cannot fix the meaning of his own texts (Prince 1984, 52; Krantz and Mandelbrot 1989, 17).[8]

Mandelbrot's awareness of his own intergeneric approach is one example of the slippage from readerly intention to writerly reality. Mandelbrot was clearly aware of the narrative ramifications of his work, as his comments in the section titled "This Essay Mixes Styles; Is Semipopular *and* Scholarly" indicate. He further assures us that "the main purpose of this Essay, however, lies elsewhere. As has been stressed in the introductory paragraphs, this work is above all a description (both monographic and synthetic) of theories and theses which I believe to be new" (1977, 23).

Mandelbrot makes the inevitable move toward closure, toward the definitive (i.e., Mandelbrotian) definition of terms. Finally, in a move consistent with his other rhetorical gestures, Mandelbrot includes part of a review of *Fractals* by Freeman Dyson (1978) in his revision, *Fractal Geometry*. Mandelbrot describes the excerpt as "an eloquent summary" (1983, 3) of the mathematics leading up to fractal geometry. He performs this summary himself in various places in the book, so this excerpt seems to serve a purpose other than providing background. The review's quoted text adds little to the reader's understanding of the material; however, it provides instant support for Mandelbrot's theories and puts them in a scientific context not unlike one created in the peer review process.

Turning to Prigogine and Stengers's text, Porush's comments might lead us to believe that it is significantly different from Mandelbrot's in both its assumptions and its methodology, yet it falls into some of the same traps as Mandelbrot's text does. Like the traditional textbook, *Order out of Chaos* starts with a history; rather than this historical account occupying the first short chapter, however, it takes up two-thirds of the book. Prigogine and Stengers contend that,

> to appreciate the reconceptualization of physics taking place today, we must put it in proper historical perspective. The history of science is far from being a linear unfolding that corresponds to a series of successive approximations toward some intrinsic truth. . . . We have tried to place the history of science in the frame of the history of ideas to integrate it in the evolution of Western culture during the past three centuries. Only in this way can we appreciate the unique moment in which we are presently living. (1984, xxviii–xxix)

This is the *nouvelle alliance* which titled the French edition—developments in physics are conditioned by developments in the culture surrounding them. Prigogine and Stengers thus disagree with the Kuhnian model of scientific development (the paradigm shift), claiming that "we have to incorporate the complex relations between 'internal' and 'external' determinations of the production of scientific concepts" (1984, 309). While Prigogine's other popular text on the subject, *From Being to Becoming* (1980), attempts to "present to a large group of readers a simple introduction to a field that seems to me to have wide implications" (xviii), *Order out of Chaos* attempts to show what those implications are inside a historical and theoretical framework. The historical chapters of the book look at the changes in the conception of physical events, explaining how classical Newtonian physics is a science of being (characterized by static or reversible events) and contemporary sciences are sciences of becoming (concerned with the irreversibility of time). Prigogine and Stengers describe this as a move from a disenchantment with nature under the paradigm of the classical or Newtonian objective observer to a "reenchantment" with nature within modern science. Somewhat optimistically they claim that

> the natural sciences have thus rid themselves of a conception of objective reality that implied that novelty and diversity had to be denied in the name of immutable universal laws. They have rid themselves of a fascination with a rationality taken as closed and a knowledge seen as nearly achieved. They are now open to the unexpected, which they no longer define as the result of imperfect knowledge or insufficient control. (1984, 306)

Compared with Mandelbrot, Prigogine and Stengers seem to be on one end of a spectrum of rhetorical intention: Mandelbrot clamors for legitimation (within a system that previously did not want him) at one end of the spectrum, and Prigogine and Stengers try to open up the system to new ideas and methods at the other. Prigogine and Stengers seem, therefore, to be embracing the writerly, that discourse which does not attempt to close down meaning, to fix it in one particular way. Aware of aspects of theory-ladenness, they attempt to make all of their assumptions obvious, available to the reader. In contrast to the revising activities of Mandelbrot, for example, Prigogine moves from a fairly traditional popularization aimed at an "intermediate" level audience in *From Being to Becoming* to one written in collaboration with a historian of science in *Order out of Chaos*. The obvious sense in the latter that science does not occur within a cultural vacuum is definitely a step toward opening the text, showing that its conclusions are influenced by forces internal and external to science itself. The historical chapters are not, then, filler or introductory material, but are as necessary a framing device as an explanation of the mathematics used in studying dissipative structures would be. Prigogine and Stengers see the historical and scientific explanations combining to explain particular developments in the physical sciences in our century: "how can we consider as accidental that the rediscovery of time in physics is occurring at a time of extreme acceleration in human history? Cultural context cannot be the complete answer, but it cannot be denied either" (1984, 309).

Despite their frequent protestations about the way science has previously been portrayed, however, Prigogine and Stengers are not completely free of strong attempts toward closing meaning. They claim that "our vision of nature is undergoing a radical change toward the multiple, the temporal, and the complex"; yet they then immediately assert that "a new unity is emerging: irreversibility is a source of order at all levels. Irreversibility is the mechanism that brings order out of chaos" (1984, 292). Dissipative structures suggest, therefore, a new grand unified theory, one that links the microscopic findings of quantum mechanics to the macroscopic manifestations of chaotic systems. The subtitle of the book proclaims that *Order out of Chaos* is "man's new dialogue with nature"; yet in it we see that the dialogue is not necessarily new, only Prigogine's version is. In reference to the initial growth of science, Prigogine and Stengers comment that

> science initiated a successful dialogue with nature. On the other hand, the first outcome of this dialogue was the discovery of a silent world. This is the paradox of classical science. It revealed to men a dead, passive nature, a nature that behaves as an automaton which, once programmed, continues to follow the rules inscribed in the program. In this sense the dialogue with nature isolated man from nature instead of bringing him closer to it. (1984, 6)

Today, they contend, "physicists have no privilege whatsoever to any kind of extraterritoriality. As scientists they belong to their culture, to which, in their turn, they make an essential contribution. . . . It is this conception of knowledge as both objective and participatory which we have explored through this book" (1984, 299).

Yet, awareness of social context is not dialogue. Even if an author actively sought to create a writerly text, certain associations and constructions of meaning would be unanticipated, would be created at the site of the reader (there is, in other words, no readerly-writerly text, no text that can control the writerly practice of the reader). Prigogine and Stengers, while describing a science of "becoming," phrase it in terms of "being." The unification metaphors (not the least of which is "order out of chaos") demonstrate to the semipopular audience that the days of theories that made nature uncommunicative are over; they have been replaced by an equally dominating (i.e., readerly) attempt at describing a now-fluid nature.[9]

It is not enough, it seems, to recognize the way scientific discursive strategies attempt to close meaning and fail; the writer aware of the constructed nature of scientific knowledge can still, like Prigogine and Stengers, become caught up in the teleological metaphor of "a new unity." Mandelbrot's text is perhaps a more common example of the textbook mentality that Kuhn describes, but we see that it shares many traits with *Order out of Chaos*. Prigogine and Stengers undermine their attempt to show that a new view of time implied by a study of "dissipative structures" carries "ethical responsibility" (1984, 312) through their failure to allow their format to echo the openness of their topic. While we might not be surprised to see the readerly in the traditional scientific text undermined through a rhetorical critique, a study of the discourse of the semipopular scientific text demonstrates that even in a text devoted to multiple discourses and suggestive language, the readerly nevertheless dominates the scientific essayist's intention, and the writerly continues to undermine it.

NOTES

1. Though I do not intend to create a taxonomy of scientific texts in this essay, it is useful to define the "semipopular" scientific text apart both from traditional popularizations and from textbooks. It seeks to be true to the science it is portraying (not simplifying it overmuch) but does skip some of the mathematical development and experimental data that would be expected in a more traditional scientific paper. It makes assertions about a science and about its implications without rigorous proof and is therefore close in form to the scientific textbook; it teaches all that is necessary and gives enough information to make that knowledge useful.

The semipopular scientific essay would include Mandelbrot's and Prigogine and Stengers's texts, and certainly some of the writings of Douglas Hofstadter, but would probably not include more traditional popularizations such as *Chaos*, by James Gleick (1987), or the popularizations by Stephen Jay Gould or Lewis Thomas. Though I cannot consider the possible reasons for it here, a look at the kinds of texts that fall into the semipopular scientific category indicates that certain subdisciplines find this form most appropriate: the chaos sciences, cognitive science, and probably some others that are interdisciplinary in their methods.

2. A classic example of this kind of scientific narrative, as described by Gross and others before him, is James Watson's *Double Helix* (1980), in which the "race for the double helix" is played out as a suspense story.

3. Though Kuhn's (1970) theories of paradigms have been challenged, his observations about the status of textbooks in the scientific community are quite insightful and still valuable.

4. In describing the degree to which scientific observations are theory-laden, Norwood Hanson suggests in *Patterns of Discovery* (1958) that even the same

visual phenomenon is perceived differently depending on the observer's theoretical background. He constructs a thought experiment in which he imagines Johannes Kepler "on a hill watching the dawn. With him is Tycho Brahe. Kepler regarded the sun as fixed: it was the earth that moved. But Tycho followed Ptolemy and Aristotle in this much at least: the earth was fixed and all other celestial bodies moved around it. *Do Kepler and Tycho see the same thing in the east at dawn?*" (5) Hanson claims that they do not, that the observation means totally different things to the two astronomers depending on their respective worldviews.

5. Werner Karl Heisenberg's uncertainty principle is probably the best-known articulation of this principle—the observer can know only so much about a sub-atomic particle, and in fact determines certain attributes to be indeterminate by the very act of measuring others.

6. Niels Bohr obviously understood the almost tautological nature of these observations when he claimed that "if anybody says he can think about quantum problems without getting giddy, that only shows that he has not understood the first thing about them" (quoted in Gregory 1988, 200).

7. H. O. Peitgen's *Beauty of Fractals* (1986) is the prime example of this phenomenon. The book is a large-format text, full of glossy photos of chaotic systems, so the mathematical formulas interspersed between the pictures do almost nothing to spoil the impression that this is a coffee-table art book.

8. Krantz and Mandelbrot's article (1989) is actually a combination of three texts—Krantz's review of Peitgen's *Beauty of Fractals* and another Peitgen collection, a retort by Mandelbrot, and a last word by Krantz. Mandelbrot saw a copy of the review and was disturbed by the fact that he was not given proper credit in the history of chaos studies, and his reply engendered a debate which resulted in the review's rejection by the *Bulletin of the American Mathematical Society*. The whole exchange was subsequently picked up by the *Mathematical Intelligencer*.

9. A more appropriate "dialogue with nature" might take the form of Evelyn Fox Keller's portrayal of Barbara McClintock in *A Feeling for the Organism* (1983). Keller's reconstruction of McClintock's sometimes neglected achievements in plant biology bring her to claim (in language not unlike Prigogine and Stengers's) that "even in those subjects that lend themselves most readily to quantification, discourse depends heavily on conventions and interpretation" (146). However, Keller's own re-presentation is an attempt to be true to the subject she has undertaken. She does not conclude with huge metaphysical assumptions about the true geometric form of nature (as Mandelbrot does) or about a "source of order at all levels" of nature (as Prigogine and Stengers do); instead she studies one example in detail (as McClintock herself did). This kind of local analysis is suitable for the study of a plant biologist who herself was interested in the isolated case that gave small insights into the whole in a gestalt way. We can contrast this with the traditional form of scientific theories, in which the "covering law" is applied to all cases. Keller's text emulates the science she studies—it moves not toward conclusions but

rather toward implications of the kind of science McClintock performed. Keller's text should serve as an example, therefore, of how to avoid attempting to impose readerly order on a chaotically writerly text.

REFERENCES

Barthes, R. 1974. *S/Z: An Essay.* Trans. R. Miller. New York: Noonday/Farrar, Straus, and Giroux.

———. 1975. *The Pleasure of the Text.* Trans. R. Miller. New York: Noonday/ Farrar, Straus, and Giroux.

———. 1986. The Reality Effect. Trans. R. Howard. In *The Rustle of Language.* New York: Hill and Wang.

Dyson, F. 1978. Characterizing Irregularity. Review of *Fractals: Form, Chance, and Dimension,* by Benoit Mandelbrot. *Science* 200: 677–78.

Gleick, J. 1987. *Chaos: Making a New Science.* New York: Viking Press.

Gregory, B. 1988. *Inventing Reality: Physics as Language.* New York: John Wiley and Sons.

Gross, A. G. 1990. *The Rhetoric of Science.* Cambridge: Harvard Univ. Press.

Hanson, N. 1958. *Patterns of Discovery.* Cambridge: Cambridge Univ. Press.

Hayles, N. K. 1990. *Chaos Bound: Orderly Disorder in Contemporary Literature and Science.* Ithaca: Cornell Univ. Press.

Keller, E. F. 1983. *A Feeling for the Organism: The Life and Work of Barbara McClintock.* New York: W. H. Freeman.

Krantz, S. G., and B. Mandelbrot. 1989. Opinion. *The Mathematical Intelligencer* 11(4): 12–19.

Kuhn, T. S. 1970. *The Structure of Scientific Revolutions.* 2d ed. Chicago: Univ. of Chicago Press.

MacCabe, C. 1985. Realism and the Cinema: Notes on Some Brechtian Theses. In *Theoretical Essays: Film, Linguistics, Literature.* Manchester: Manchester Univ. Press.

Mandelbrot, B. 1975. *Les objets fractals: forme, hasard et dimension.* Paris: Flammarion.

———. 1977. *Fractals: Form, Chance, and Dimension.* New York: W. H. Freeman.

———. 1983. *The Fractal Geometry of Nature.* New York: W. H. Freeman.

Peitgen, H. O. 1986. *The Beauty of Fractals: Images of Complex Dynamical Systems.* Berlin: Springer-Verlag.

Porush, D. 1990. Making Chaos: Two Views of a New Science. *New England Review and Bread Loaf Quarterly* 12(4): 427–42.

Prigogine, I. 1980. *From Being to Becoming: Time and Complexity in the Physical Sciences.* New York: W. H. Freeman.

Prigogine, I., and I. Stengers. 1979. *La Nouvelle Alliance.* Paris: Gallimard.

————. 1984. *Order out of Chaos: Man's New Dialogue with Nature*. Toronto: Bantam Books.

Prince, S. D. 1984. In the Mind of Dr. Benoit Mandelbrot. *Computer Pictures* 2(3): 46–52.

Watson, J. 1980. *The Double Helix: A Personal Account of the Discovery of the Structure of DNA*. Ed. G. Stent. New York: W. W. Norton.

White, H. 1978. The Historical Text as Literary Artifact. In *Tropics of Discourse: Essays in Cultural Criticism*. Baltimore: Johns Hopkins Univ. Press.

Making Chaos: Two Views of a New Science

DAVID S. PORUSH

The world may indeed be considered as a vast Machine, in which the great Wheels are originally set in Motion by those which are very minute, and almost imperceptible to any but the strongest Eyes.
—Henry Fielding, *Tom Jones*

James Gleick's best-selling book *Chaos* (1987) is a fascinating, well-paced, extraordinarily absorbing, and persuasive account of "the making of a new science." *Chaos* not only brought the arcane and technical material of an important scientific enterprise to a much wider audience, from Wall Street to Hollywood, it also alerted scientists and academics in other disciplines to the fundamental shifts in worldview that chaos science implies. Perhaps it even drew scientists from other fields to look seriously at the power of nonlinear mathematics to model intractable phenomena— one of the essentials of chaos theory—thereby fostering the sort of interdisciplinarity Gleick so praises in his book.

Like many science popularizations, *Chaos* consciously uses narrative tools and techniques borrowed from fiction. Take, for instance, the opening lines of chapter 1, which parody the pulp western: "The sun beat down through a sky that had never seen clouds." We find in the next sentences, though, that we are in a peculiarly postmodern setting: "The winds swept across an earth as smooth as glass. Night never came, and autumn never gave way to winter. It never rained." Gleick has ingeniously located us in that most postmodern of spaces, a fitting locus for this cybernetic age: inside a computer simulation. "The simulated weather in Edward Lorenz's new electronic computer changed slowly but certainly, drifting through a permanent dry midday midseason, as if the world had turned into Camelot, or some particularly bland version of southern California" (11).

CHAOS AS POSTMODERN MYTHOLOGY

As the French sociologist Jean Baudrillard has suggested, the hallmark of our era is that the simulation has displaced reality, the map has supplanted the original. We abide not in any home granted us by nature but in the blueprinted structures of representation and invention. And Gleick's tale does in some respects read as a sort of postmodern mythology. Not unlike postmodern novels by John Barth, Thomas Pynchon, William Burroughs, and Kurt Vonnegut, *Chaos* is distrustful of orthodox constructions of knowledge, of science's prevailing mythologies and paradigms. It is concerned with how people "really" know, as opposed to *how we are told to know* by official versions of reality. Furthermore, like most postmodern fictions, *Chaos* self-consciously encodes its narrative according to a theory of how narratives tell. In *Chaos* the hero is not a character but an idea, or more properly a group of interrelated ideas, and Gleick fits the events of his story to a pattern rather than letting the pattern emerge from events.

For instance, Gleick's narrative of the course of this new heroic idea on the scene in science in some respects follows Joseph Campbell's cycle of the heroic journey. Chaos, like Beowulf, "emerged from a backwater" to challenge "the glittering abstractions of high energy particles and quantum mechanics," giving hope to "younger physicists" who felt "the field had been dominated long enough" by those ideas (Gleick 1987, 6). Such an analysis explains the occasional hyperbole of *Chaos*: "The most passionate advocates of the new science go so far as to say that twentieth-century science will be remembered for three things: relativity, quantum mechanics and chaos. Chaos they [many physicists] contend has become the century's third great revolution in the physical sciences. Like the first two revolutions, chaos cuts away at the tenets of Newton's physics" (6). Most of Gleick's book is intended to explain why chaos deserves such passionate advocacy even as Gleick sketches with remarkably lucidity the shape of chaos itself.

But an even more fitting pattern for this tale lies less far afield than Campbell's model: Gleick tells his readers, though not in so many words, that he is a devout Kuhnian. He clearly sees in the "new science" of chaos a fulfillment of the pattern for dramatic scientific change detailed (and one might say prophesied) in Thomas Kuhn's landmark book about scientific progress, *The Structure of Scientific Revolutions* (1970). In fact, the first few pages of the second chapter of *Chaos*, entitled "Revolutions," are devoted to an explicit discussion of Kuhn's version of the history of science. At first glance, such a discussion has no direct bearing on the story of chaos. Rather, like a good postmodern narrative, it seems to be a self-conscious inclusion of the author's model for his tale in the tale itself. However, one broad message of postmodernism is that such inclusions are not neutral or value free. In a post-Heisenbergian world, there is no such thing as an epistemologically neutral point of view.

In chapter 2 of *Chaos*, Gleick ticks off the points in which the progress of chaos studies through the 1970s and early 1980s followed Kuhn's scheme: the field was dominated by safe, conservative, modest, "orthodox" science; chaos, by contrast, was "the exceptional, unorthodox work that creates revolutions" (36). Gleick is critical of "normal science," which "consists largely of mopping-up operations" designed by scientists who have been indoctrinated to attend to mere "puzzles" in the delusion that they were treating significant problems (36). But he lauds the idealism of a revolutionary science that crosses disciplinary boundaries and endured early scorn, suppression, dismissal, discouragement, or open hostility. In passages worthy of Tom Wolfe, Gleick even takes pains to paint the personal flair of the new breed of avant-garde scientists. He examines how results or speculations were stylistically shaped and reported in journals, and how reporting chaos in unconventional terms was virtually de rigueur. Early papers "sounded evangelical. . . . They declared new credos and they often ended with pleas for action" (39). The successful revolution ushers in a new age. Accordingly, by this strictly Kuhnian interpretation, "chaos has become not just a theory but also a method, not just a canon of beliefs but a way of doing science" (38). Finally, in what I believe is the most telling of these Kuhnian points echoed in *Chaos*, Gleick writes: "New hopes, new styles, and, most important, a new way of seeing. Revolutions do not come piecemeal. One account of nature replaces another. . . . In Kuhn's words, 'It is rather as if the professional community had suddenly been transported to another planet where familiar objects are seen in a different light and are joined by unfamiliar ones as well'" (39). This idea—that paradigms are mutually exclusive perceptions of the world, that one view of nature "replaces" another—is certainly the most radical of Kuhn's notions, and one that explains a good deal of the drama and style of Gleick's report, for there is something elementally appealing about such Oedipal displacements, even when applied to scientific ideas.

IS THIS THE ONLY WAY TO TELL IT?

Yet, one wonders if there aren't other accounts of the rise of this new science just as revolutionary or as sensational. Is it possible that in adopting the Kuhnian scheme as the pattern for his reportage, Gleick has fashioned a patent *fiction* of how the so-called science of chaos came out of the neglected backwaters, the "wastebasket science" of physics, against all odds, to dominate the mainstream, overthrow the ruling paradigm, and usher in a new age?[1] The word *fiction* doesn't necessarily imply malicious distortions or deliberate falsifications, merely tacit or inevitable ones. And so when Gleick adopts whole cloth an analytical pattern for seeing recent science history that fits Kuhn's scheme, it inevitably excludes other competitive, and perhaps even more valid, versions. After all, to translate is to interpret. Or as the Italians say, *traduttore, trattore* (translator, traitor). It is naïve to think that such an exclusive interpretation is merely an idle, passive recording of events rather than an active force in the history of science. Perhaps the undeniable power and attractiveness of Gleick's tale have helped change the direction of chaos science itself. Perhaps it unconsciously promoted some lines of research and ignored others. Perhaps by fixing the character of chaos in the cultural imagination in a certain way, it enticed certain firebrand younger scientists and deflected others. In fact, ironically, understanding that representations of science help shape science itself is one of the fundamental lessons of Kuhn's work. Paradigms precede and lead perceptions; theories organize the facts that support them and suppress those that don't. Facts don't exist without the structure of knowledge—the context, the narrative—that supports them.

But Kuhn's meta-paradigm, his model or paradigm of how change occurs in science through paradigms, is not the only pattern for scientific progress. Persuasive supplementary or alternative constructions of science have been offered by structuralist Michel Foucault (1972), anarchist-historicist Paul Feyerabend (1987), cosmologist Stephen Toulmin (1982), neorationalist Imre Lakatos (1978), and feminist Evelyn Fox Keller (1985), to name just a few. These philosophers and historians of science suggest certain shortcomings in Kuhn's view, tacitly or explicitly: Kuhn does not sufficiently appreciate language or question the relationship between academia and popular constructions of science; his view seems deterministic; it relativizes everything; it remains silent on issues of gender, class, culture, and race; it is too rational; it is not rational enough.

WHO MADE CHAOS?

When one stops to think, then, that Gleick's Kuhnian narrative is only one possible version of how science is made, the word *making* in the subtitle of the book obtains a lurking irony. Where is the science "made" and by whom? Clearly, Gleick intended us to understand how some heroic actors—Mitchell Feigenbaum, Edward Lorenz, Benoit Mandelbrot, Steve Smale, and others—in the grips of an even more heroic set of concepts, prevailed against the odds to make "chaos" and usurp established paradigms. But if in this century we've learned one thing that crosses all disciplines to influence science and criticism and social science and philosophy, it is that discourse shapes the subject. *The observer's point of view shapes the observed*. Gleick himself, as much as or even more than his heroic scientists, has "made" the science, not least by choosing a very particular model or theory of how science works. Thus the title, though unintentionally, comments on itself.

The degree to which Gleick has successfully "made" chaos, or perhaps we might claim more soberly, helped to make chaos, could be measured by answering a set of questions such as, What sort of new lines of research have become popular among scientists after the publication of the book? Who has been directly attracted into the field by reading the book? What lines of research have suffered or withered away because they are neglected in the book? How effective has *Chaos* been in persuading other journalists that Gleick's shape of the science is the science? All these questions are interesting, but here I am going to pursue another line, one that strikes at serious matters about the relationship between telling and doing in science, and to the heart of questions of how one comes to understand the world through physics. My two questions are these: What alternative views of this field of scientific study have been effectively suppressed or chilled by the success of Gleick's version? Are there important implications of the actual discoveries included under the rubric of "chaos" that Gleick fails to recognize or include as the result of his particular view?

AN ALTERNATIVE VIEW OF CHAOS

Ilya Prigogine won the 1977 Nobel Prize in chemistry for describing mathematically a general class of phenomena he calls "dissipative structures." Prigogine has popularized his own work in *Order out of*

Chaos (1984), a notable book about cosmology and science philosophy co-authored with the historian of science Isabelle Stengers. As the very title implies, Prigogine's interest and the work for which he was recognized are closely related to the science of chaos that Gleick studied.

Prigogine's theory of dissipative structures suggests that ordered systems arise spontaneously out of conditions that look chaotic but really harbor a hidden ordering principle or potential. His mathematics predict when and tell why certain kinds of unstable systems will leap to order in surprising, almost intelligent-looking, ways. According to Prigogine, this self-organization out of chaos is actually *likely* to occur under the conditions he stipulates, and, even more surprising, these conditions are actually quite common in nature, particularly in chemical and biological systems.[2]

In short, Prigogine's powerful work provides a model for the rise of complex, self-organizing systems from apparently chaotic conditions. But Prigogine's work was recognized not only for its description of this process of self-organization, but also because of the implications of his model for reconciling the hitherto irreconcilable worldviews of entropy and evolution. Until Prigogine's work was recognized, it was difficult to explain how the teeming richness of life on earth arose on its own—seething with complex order, proliferating specializations, and growing complexity—while the second law of thermodynamics mandates that the universe is winding down into an inevitable, entropic heat death, dispersing organized structures until they achieve a maximized simple randomness.

In the same way, until Prigogine, it was hard to reconcile time as physics views it with time as humans experience it. In the perfected Newtonian model, collisions of billiard balls work forward or in reverse: time is *symmetrical*. Yet, all living things clearly experience time moving in one direction, *asymmetrically*. We all know that to live is to move inexorably toward extinction.

Finally, until Prigogine, it was hard to understand how science remained content with offering two levels of description of everyday phenomena. On one hand, there were mathematical descriptions of natural events which presumed a kind of ideal world in which messy (nonlinear) events like friction had to be discounted in order for scientists to make sense of things. On the other hand, there were the more narrative and messy tactics needed to describe complex biological systems.

In short, the territory Prigogine explores is virtually the same territory explored by Gleick and his subjects, so much so that one could safely say that the same "facts" are being explained. However, they are reconstituted from the perspective of a different version of science's role, and thus the two explanations come to strikingly different conclusions and portray very different applications.

With this in mind, then, it is both shocking and yet somehow unsurprising that Prigogine's work is never mentioned in the pages of *Chaos*. Prigogine's name occurs only once, in an oblique reference buried in a footnote to one of the last pages of the book. There, Gleick talks about "thoughtful physicists" who have reconsidered the role of entropy and the second law of thermodynamics in cosmology. The note to this comment reveals that Gleick isn't even thinking about Prigogine directly, but rather about P. W. Atkins, whose 1984 book, *The Second Law*, Gleick greatly admires for its discussion "of the creative power of dissipation in chaotic systems." Then, and only then, Gleick also mentions Prigogine in terms that, were science popularization a cocktail party, would seem like a snub: "A highly individual, philosophical view of the relationships between thermodynamics and dynamical systems is Ilya Prigogine's *Order out of Chaos: Man's New Dialogue with Nature* (New York: Bantam, 1984)" (339).

This citation is peculiar on several counts. First, despite the obvious relation of the title to the concerns of Gleick's book, in this one-clause gloss of the book there is no hint that Prigogine's work also explores the revolutionary role of nonlinear models in explaining how order arises out of complexity. Furthermore, there is no hint of the impact Prigogine's work has had on chaos studies in ways that would fulfill even Gleick's Kuhnian requirements for a revolution. For instance, Prigogine's thinking has promoted highly original, interdisciplinary work in biology, ecology, traffic studies, astrophysics, thermodynamics, neurology, biophysics, chemistry, particle physics, and even the social sciences and management. The Ilya Prigogine Center for the Study of Statistical Thermodynamics has become a powerful force on the University of Texas, Austin, campus, where colleagues from many disciplines have collaborated on applying the dissipative structure model to various phenomena. Prigogine has even influenced the popular imagination. Science fiction novels by Bruce Sterling (1977, 1985, 1988, 1989), Lewis Shiner (1988), and A. A. Attanasio (1981) explicitly apply Prigogine's theories to their own speculative portraits of the future. There is, of course, no mention by Gleick of Prigogine's Nobel Prize. Finally, and most tellingly, the phrase "highly individual, philosophical view" is tantamount to excommunication from the church of science (even after the Kuhnian reformation), where canons of objectivity and impersonality are still hostile to the "highly individualistic and philosophical." Gleick, despite his own Kuhnian praise for other excommunicants who were once treated as cranks, and however unintentionally, makes Prigogine sound like a bit of a crank himself.

How can we explain this remarkable confinement of Nobel Prize–winning work in chaos dynamics to one dismissive appendage in an obscure footnote?

TO THE SOURCES THEMSELVES

As I was escorting him from his talk to a reception in his honor at a conference in October 1988,[3] I stole the opportunity of asking Prigogine himself how he viewed the matter. A short, stocky man, well into his seventies at that time, Prigogine moved and spoke with deliberateness. His heavy Belgian-Russian accent gave his English an even more circumspect quality, and his Old World manners made me hesitate to ask so potentially rude a question. Not surprisingly, he seemed quite alert to the issue of his absence from the pages of Gleick's book. I had long considered asking him the question and had anticipated all sorts of answers: that his work was really in statistical thermodynamics or in chemistry and Gleick was concerned more with physics, computer modeling, and mathematics; or that there were nontrivial but arcane differences between the theoretical approaches to similar facts; or that there were disagreements over the applicability of the mathematics to particular phenomena. Or, most generously, Gleick simply did not understand Prigogine's work.

None of these, I have since learned, is true, though all have hints of the truth in them. And none of these suppositions even came close to predicting Prigogine's blunt reply: "Well, I suppose it's because he [Gleick] was so interested in Americans, you see, and not Europeans." It astonished me to think that Prigogine would attribute Gleick's omissions to American chauvinism, and then, even more, to think that he might be right. So I went back to explore the territory itself.

The book *Chaos* contains surprisingly little on Europeans' work in the field. While some of the heroes of the book are European by birth (Mandelbrot, for instance), they did most of their work in the United States (Mandelbrot for IBM). And generally, one gets the impression that chaos is primarily an American enterprise: of the twenty or so institutes that Gleick acknowledges in his notes, only two are European, and of the hundred or so personal acknowledgments, only three or four are of scientists working in Europe.

Yet, Europeans were intensely involved in work on all the primary aspects of chaos on which Gleick focuses: bifurcation, fractals, strange attractors, nonlinear systems. Indeed, a 1981 volume of papers presented at an international conference in West Germany, *Chaos and Order in Nature*, includes vanguard papers by scientists from West Germany, East Germany, The Netherlands, France, the Soviet Union, the United States, Great Britain, Austria, and Sweden (Hakan, 1981).[4] To be fair, Gleick does not particularly describe chaos as the sole province of American researchers and shows no self-evident suppression of foreign work in the field. I have tried, therefore, to take the most generous position on this curiosity. I cannot be persuaded that Gleick's tactic is malicious or chauvinistic, but neither can I be persuaded that it is simply unconscious, for Gleick began working as a science reporter for the *New York Times* when the impact of Prigogine's work was still being recounted in its pages (in 1978). So I decided to ask Gleick himself and phoned him at his home in Brooklyn.[5] I asked him why Europeans in general, and Prigogine in particular, were absent from his pages.

His response was striking. He first began by ticking off the names of Europeans that do occur in his book: David Ruelle, Daniel Henon, Brosl Hasslacher. But then he challenged me by asking me where *I* would put Prigogine in the book. The challenge is a clever one. If Gleick really does have a very different view of how science progresses from the one Prigogine espouses, then according to the laws of Kuhnian perception, those views are very likely to be mutually exclusive. There is no place to fit a wide-angle view into a film shot entirely with a telephoto lens. As Gleick put it, he was "trying to tell a particular kind of story. . . . The scientists I talked to, who were part of this story, were not thinking of Prigogine," he said. "They maybe were thinking of Goethe even," he laughed, "and Smale, but not Prigogine. I was just trying to follow the story of what they were thinking of. . . . Even if I wrote the book again, I wouldn't know where to put Prigogine. And anyway, by 1980, that story was over."

I tried another tack. I asked him if he didn't think that the science Prigogine explored bore some relation to the science explored by his subjects. And here is where that exclusive perception problem became most pronounced. He responded with a self-effacing, disarmingly modest, and yet revealing perspective: "Let me put it this way. I think that when people think about the field of chaos and look back on it, they're not going to mention Prigogine's name, any more than they're going to mention my name."

This remarkable underestimation of his own impact on the field is in keeping with an important Kuhnian point of view: despite its vulnerability to cultural and rhetorical considerations, science is still an academic, professional enterprise conducted by scientists. Popularizations, even rigorously reported ones like Gleick's, fall outside the bounds of the great institution of "science." T. H. Huxley's retelling—or retailing—of Darwinism secured Darwin's fame, not his own. In other words, despite the undeniable impact of his book, Gleick discounts his own role in "making" chaos, though years from now, people will undoubtedly mention Gleick when they mention chaos. In this postmodern era, the teller and the told are more inextricably bound than ever before. Gleick's telling of chaos, in many minds, made his name synonymous with the science. And if years from now the other half of Gleick's prophesy comes true and people don't mention Prigogine's name when they think of chaos, then that might well be a result of Gleick's journalism, too. Nonetheless, there is no denying the direct relationship between the work of chaoticians and Prigogine's Nobel Prize–winning investigation on many levels, from the mathematics it invokes to the range of phenomena it seeks to explain to its interdisciplinary expression. Somehow, Prigogine fell outside Gleick's loop, and this cannot be explained merely as a case of Gleick's patriotism.

In short, the testimonies of both principals are unsatisfactory. Prigogine feels he belongs in those pages and blames it on mere chauvinism; Gleick feels Prigogine's work—the part he's familiar with, anyway—just doesn't play a role in the story he was telling.

TWO VIEWS OF SCIENCE ITSELF

A third, deeper explanation for this curious sin of omission lurks elsewhere, deeply hidden in the most fundamental commitment by two authors who are providing competing explanations of a breakthrough not only in scientific fact but in the scientific worldview. To see this, we must turn to Prigogine not as the Nobel Prize–winning scientist but as Gleick (and many other scientists) sees him, as the coauthor of a quite speculative popularization of his own work.

A little like those athletes who hire an author to help write their views from the locker room, Prigogine teamed up with Isabelle Stengers to delve some of the more far-flung social and philosophical implications of his work. In one of the most fascinating passages in *Order out of Chaos*, Prigogine applies his own model of how order arises out of chaos to the emergence of a new cosmology in science out of what Kuhn would call "normal," and Gleick "orthodox," research. It comes on one of the last pages, where Prigogine is considering how his own evolutionary view of complex order might serve as a model for scientific progress, in contrast to Kuhn's revolutionary model. Prigogine explains his primary and direct experience in the making of dissipative structure theory, and at the same time he tries to apply that theory to develop *a richer view of science as a cultural phenomenon*. In other words, Prigogine applies his own view of order arising *naturally* out of *apparent* chaos to explain how science progresses:

> Here we must disagree somewhat with Thomas Kuhn's analysis of the formation of "normal" science. . . . In Kuhn's view, the transformation of a paradigm appears as a crisis. . . . When nature is eventually seen as refusing to express itself in the accepted language [of the ruling paradigm], the crisis explodes with a kind of violence that results from a breach of confidence. At this stage, all intellectual resources are concentrated on the search for a new language. Thus scientists have to deal with crises imposed upon them against their will.
>
> The questions we have investigated have led us to emphasize aspects that differ considerably from those to which Kuhn's description applies. We have dwelled on continuities, not the "obvious" continuities, but the hidden ones, those involving difficult questions rejected by many as illegitimate or false but that keep coming back generation after generation. . . . In fact, the interest of such questions is hardly surprising. To us the problem is rather to understand how they could ever have been neglected after the work of [novelists like] Diderot, Stahl, Venel. . . . [My discoveries are] not to be seen as some kind of "revelation," the possession of which would set its possessor apart from the cultural world he lives in. On the contrary, this development clearly reflects both the internal logic of science and the cultural and social context of our time. (Prigogine and Stengers 1984, 307–9)

In other words, Prigogine, like Gleick, has also self-consciously applied a narrative scheme to his framing of the tale of a new science. But Prigogine's model, by definition, is richer than Kuhn's (and consequently Gleick's). Kuhn's view is nearly positivistic, more orthodox than it acknowledges, and a bit mechanical in its view of how old paradigms give way to new ones. By contrast, Prigogine offers a model of a "normal" science in which nagging questions persist—sometimes preserved by philosophers of science who are in the same culture but outside the restricted club of institutional scientists who ignore those same questions—until science evolves the "language" to deal with them. Prigogine claims in his work to have discovered that new language to deal with the nagging question of apparent chaos and its role in self-organizing systems.

Most intriguing of all, this new language of science is not exclusive to scientific inquiry. Indeed, the same questions nag novelists and scientists alike. The culture as a unity seems to embark on its project of questioning and describing nature. This is a significantly and radically different view from Kuhn's tacit portrait of an insular, institutionalized science, "best considered in the context of the contemporary university" (Prigogine and Stengers 1984, 307).

CHAOS AS MISNOMER

The consequences of Prigogine's view for our understanding of science itself are profound. Adopting the more restrictive view of science as a mechanism of knowledge seeking set apart from other parts of our culture almost inevitably leads to a more restrictive view of the philosophical and cultural implications of science. This is precisely, I would argue, the problem with Gleick's accounts of chaos: it is inspiring, exciting, dramatic, but philosophically impoverished.

Perhaps the best way to illustrate this is metonymically. Contrasting the words the two writers chose to name the field is a study in miniature of this philosophical problem. Gleick calls the science "chaos," following leads provided by the scientists he spoke to. Prigogine calls the phenomena the science studies "dissipative structures," by which he intended to describe the dynamic evolution of structures that feed off the general tendency toward dissipation in the universe.

Yet even by its own terms, what emerges from any broad inspection of what Gleick calls chaos is a very simple and surprising revelation: *the word chaos is a misnomer*! Indeed, in many of the journal articles and reports about the discoveries and applications of Feigenbaum numbers, fractals, and the butterfly effect that predate Gleick's book, the proper name is "deterministic chaos."

The true revelation of chaos studies described by Gleick is not that order arises out of chaos, but that *some systems that are apparently chaotic are actually just complex (nonlinear) systems waiting for proper conditions to express a hidden order.* In other words, this so-called revolution actually reasserts one of the fundamental axioms of science: the universe can be described in deterministic mathematics. That is why so many scientists, engineers, stockbrokers, and meteorologists—anyone who has a stake in finding the apparently chaotic world a rational place that will yield to mathematical analysis—find Gleick's mythology so comforting. And indeed, the field of chaos studies in America is dominated by mechanistic applications. The concluding glory and centerpiece of the chaos revolution as told by Gleick is an enhancement of our computer modeling. Now we can program computers to give even better models of what occurs in formerly incomprehensible situations like weather prediction, fluid dynamics, population studies, and stock market fluctuations. Chaos mathematics even allows computers to generate incredibly realistic pictures of complex natural shapes like leaves, alluvial plains, and mountains, giving chaos its own glitzy promotional materials.

To put it even more simply, calling the entire science that discovers deterministic descriptions in apparent chaos "*chaos*" is tantamount to calling the Copernican revolution "geocentrism" instead of "heliocentrism." Insisting that chaos is a sort of Kuhnian revolution in which one whole worldview is overturned in favor of another simply disguises this reassertion of the deterministic orthodoxy; it is a sort of unintentional doublethink.

By contrast, Prigogine's view emphasizes natural evolution and reconciliation. Indeed, the conceptual point Prigogine stresses about his own work is its insertion of irreversibility—the human element of time—into the cosmological picture. Prigoginean dynamics reconciles physics with biology and philosophy, portraying a universe in which evolutionary principles are at work at all levels. Most intriguing of all, however, is the unity of this vision. Prigogine's own discoveries in science have led him to appreciate the role of the human point of view and human discourse in shaping science, a factor to which Gleick proves himself deaf when he denies his own impact on chaos studies.

"Classical science aimed at a 'transparent' view of the physical universe," Prigogine writes. The postmodern trend, as compared with the classical one, leads to a kind of "opacity," in which words themselves, discourse and point of view, influence the facts. Prigogine quotes philosopher Herman Weyl to express this view: "Scientists would be wrong to ignore the fact that theoretical construction is not the only approach to the phenomena of life; another way, that of understanding from within (interpretation), is open to us" (Prigogine and Stengers 1984, 311).

In Gleick's hands, the final effect of the so-called revolution of chaos is a triumph for mathematics and technology: we can now get the computer to simulate hitherto intransigent phenomena with some very pretty pictures. The opaque has been made transparent: chaos presents a new way of reasserting determinism. But in Prigogine's hands, the discovery of order evolving from chaos carves a role for the personal view, for interpretation, even for the unconscious and for creativity—activities that "break the temporal symmetry of the object," so that, as Prigogine says, out of "the noise in which we live arises music" (Prigogine and Stengers 1984, 312). In other words, Prigogine's theory of chaos entails a new way of seeing and describing and a new appreciation for human cultural acts in a cosmos that now seems much less like an alien landscape constructed in a computer under "a sky that has never seen clouds" than a more familiar, if complexly human and unpredictable, abode.

THE "NOUVELLE ALLIANCE"

In the end, then, Gleick's work, for all its revolutionary rhetoric, narrates a victory for orthodox science; yet another set of intractable phenomena—nonlinear dynamical systems—has succumbed to a deterministic model. Prigogine's narrative of this same evolution of science, by contrast, entails a philosophical and discursive revolution: though determinism is still victorious, we have had to invent a new language in order to express that victory. And this new language, according to Prigogine, is a language of complexity that carries with it a new appreciation for the descriptive power and realism—the epistemological potency—of the human point of view and discursive acts of art and narrative. There is now a place in science for common human perspectives: the languages of human time and experience.

Perhaps the simplest way to see Prigogine's point is to look at an interesting convergence between the language of science and the language of an artistic narrative. In a delightful passage from the eighteenth-century novel *Tom Jones* (1749), Henry Fielding attempts to render his view of time, chance, and causality in terms of the Newtonian-Cartesian worldview of a Clockwork Universe in which he is utterly immersed:

[Sophia] was playing one of her Father's favourite Tunes, and he was leaning on her Chair when the Muff fell over her Fingers and put her out [of her tune]. This so disconcerted the Squire that he snatched the Muff from her, and with a hearty Curse, threw it into the fire. *Sophia* instantly started up and with the utmost Eagerness, recovered it from the Flames.

Though the Incident will probably appear of little Consequence to many of our Readers, yet trifling as it was, it had so violent an Effect on poor *Jones* that we thought it our Duty to relate it. In reality, there are many little Circumstances too often omitted by injudicious Historians from which the Events of the utmost Importance arise. The World may indeed be considered as a vast Machine, in which the great Wheels are originally set in Motion by those which are very minute, and almost imperceptible to any but the strongest Eyes. (170)

Though the view of a great perfectly ordered machine is prominent in this description, the reader familiar with "chaotic" views of time will recognize a very apt description of the "butterfly effect." The butterfly effect is named after the radical idea that a butterfly flapping its wings in Tokyo could cause a tornado in the United States. In chaos theory this is somewhat more technically called "sensitive dependence on initial conditions." It means that very small wheels can ultimately set in motion great ones: for want of a nail the battle was lost. Without a recognition of the powerful role the butterfly effect plays in human destiny, Daniel DeFoe, Charles Dickens, Leo Tolstoy, Jane Austen, George Eliot, Fyodor Dostoevsky, Marcel Proust— virtually every great novelist—would have been out of business. But before chaos theory, such a view of human experience was alien to science. It was dismissed as pertinent to the realm of accident, coincidence, and kismet. Newtonian time and causality were strictly idealized, symmetrical, and formal, and operated best at the microscopic level where nonlinear effects like friction might be discounted. Reactions could be reversed, and all interactions could be reduced to very simple laws. Chaos, however, unfolds to our view a nature that seems to be speaking the same language as these great artists.

Prigogine, perhaps because he is both discoverer and reporter of this new model for how order arises from apparent chaos, delves deeply. He finds lurking in chaos theory the promise of a brand-new possibility in our culture: perhaps the language nature speaks is not so different from the language of great art and our most intimately human perceptions of destiny.

This is the promise of a grand reconciliation between the two cultures of art and science, between human experience and scientific arcana. In short, the real revolution in chaos studies is that it has begun what Prigogine calls a "new dialogue with nature."

NOTES

1. An alternative interpretation, one that is by no means unlikely, is that the scientists Gleick himself interviewed see themselves in a Kuhnian light. Whether or not they had directly read Kuhn, they were immersed in a scientific culture that has been influenced and altered by Kuhn's view. In turn, Kuhn's work was itself part of a larger revolutionary rhetoric that suffused the American educational system of the 1960s and 1970s out of which these scientists emerged. It was virtually impossible for scientists with a certain combination of talent, temperament, and politics—as well as some disappointing experiences in getting their ideas accepted—not to see themselves as romantic or revolutionary heroes fighting against an "establishment" science, given the temper of the times So, from this point of view, Gleick was in fact merely reporting a rhetoric that permeated the scientists' views rather than imposing his own.

2. Namely, far-from-equilibrium, open systems that fluctuate nonlinearly.

3. Prigogine was keynote speaker at the Second Annual Conference of the Society for Literature and Science, held in Albany, New York, sponsored by Rensselaer Polytechnic Institute, 7–9 October 1988.

4. The wording of Hakan's preface is interesting in light of his field being in transition:

At our previous meetings on synergetics, the self-organized formation of structures in quite different disciplines stood in the foreground of our interest. More recently it has turned out that phenomena characterized by the word "chaos" appear in various disciplines, and again far-reaching analogies in the behavior of quite different systems become visible. In the strict mathematical sense we are dealing here with deterministic chaos, i.e., irregular motion described by deterministic equations. While in this relatively young field of research computer experiments and computer simulations predominated in the past, there now seems to be a change of trend, namely to study certain regular features of chaos by analytical methods. (v)

5. On 18 September 1989.

REFERENCES

Atkins, P. W. 1984. *The Second Law*. New York: W. H. Freeman.

Attanasio, A. A. 1981. *Radix*. New York: William Morrow.

Feyerabend, P. 1987. *Farewell to Reason*. London and New York: Verso.

Fielding, H. [1749] 1973. *Tom Jones*. Ed. S. Baker. New York: W. W. Norton.

Foucault, M. 1972. *The Archaeology of Knowledge and the Discourse on Language*. Trans. A. M. S. Smith. New York: Harper Colophon.

Gleick, J. 1987. *Chaos: Making a New Science*. New York: Viking Press.

Hakan, H., ed. 1981. *Chaos and Order in Nature*. Berlin: Springer-Verlag.

Keller, E. F. 1985. *Reflections on Gender and Science*. New Haven: Yale Univ. Press.

Kuhn, T. 1970. *The Structure of Scientific Revolutions*. 2d ed. Chicago: Univ. of Chicago Press.

Lakatos, I. 1978. *Mathematics, Science, and Epistemology*. Cambridge: Cambridge Univ. Press.

Prigogine, I., and I. Stengers. 1984. *Order out of Chaos: Man's New Dialogue with Nature*. New York: Bantam Books.

Shiner, L. 1988. *Deserted Cities of the Heart*. New York: Doubleday.

Sterling, B. 1977. *Involution Ocean*. New York: Jove.

———. 1985. *Schismatrix*. New York: Ace Books.

———. 1988. *Islands in the Net*. New York: Ace Books.

———. 1989. *Crystal Express*. New York: Ace Books.

Toulmin, S. 1982. *The Return to Cosmology: Postmodern Science and the Theology of Nature*. Berkeley: Univ. of California Press.

Aspects of the Daemonic in Primo Levi's

Periodic Table

BRUCE CLARKE

Since its release in English translation in 1984, Primo Levi's *Periodic Table* has generated a number of responses in the United States. Some readers look at it through the lens of his other autobiographical writings, set during and immediately after World War II.[1] This group views *The Periodic Table* as "an effervescent gloss on Auschwitz in the form of meditations on various chemical elements" (Tax 1986, 11), or as a work using "the chemical elements as a bridge to weave an unusual account of his experiences in the Nazi death camps" (Tagliabue 1987). I do not mean to discount the importance of the Holocaust, in general and as part of the texture of *The Periodic Table*, but only one of its twenty-one chapters—"Cerium"—deals directly with the Holocaust. The remaining chapters contain some incisive brief vignettes of Mussolini's Italy (Milan and Turin during the late 1930s and early 1940s) from the perspective of an intellectually precocious and socially awkward Jewish adolescent.

From *The Periodic Table* it emerges that Levi personally survived Auschwitz to a large extent because he had chemical training and thus a skill he could trade for a chance at survival. *The Periodic Table* is a payment on his personal and professional debts to the discipline of chemistry, and as such it is also largely a reprieve from Holocaust topics. As Levi makes clear, the book celebrates the poetry of chemistry: "Mendeleev's Periodic Table . . . was poetry . . . and come to think of it, it even rhymed!" (41). So Levi need not be read only within the context and associations of Holocaust literature. Rather, in the same vein as Lewis Thomas, Levi can also be read as the bard of a scientific vision that transcends particular historical events.[2] But whereas the scientific and technological optimism of the American 1960s informs Thomas's reflections, Levi's voice and message are tempered by his personal immersion in the horrors of the European 1940s.

Thus it would seem equally proper to characterize *The Periodic Table* as primarily a contribution to the literature of science; as, for instance, "an extraordinary, nimble, fluent book from an extraordinary life, part autobiography, part fiction, but essentially something like a memoir of elemental matter" (West 1984). As an autobiographical memoir of a life bound up with the science of chemistry, Levi's "periodic table" is arranged not by atomic number but in the chronological order of his personal experiences with particular elements. Nicely, however, the first chapter breaks even that schema, in that "Argon" is placed as the alpha of his text not because of any particular early experience of his with that element but presumably because it begins with an *A*, and certainly because the property of inertness that argon shares with the other noble gases—an "unwillingness" to combine with other substances—reminds him of his ancestors. Here and elsewhere throughout the text, the topic of chemistry releases Levi from the mode of prosaic memoir into the mode of poesis. Even while adhering to the actual chemical qualities of elements, he frees his muse through personification of the qualities that matter and human character seem to share: the aloofness of argon, the volatility of potassium, the affability of tin, the gregariousness of carbon, and so forth. Thus, "Argon" frames a splendidly playful meditation on his immediate and extended families, and on the Piedmontese Jewish dialect by which for centuries they linguistically enforced their segregation from neighboring Gentiles.

Nevertheless, at the end of *The Periodic Table* Levi contends that it "is not a chemical treatise. . . . Nor is it an autobiography. . . . It is—or would have liked to be—a micro-history, the history of a trade and its defeats, victories and miseries" (224). Rather than simply discount such denials as authorial reticence or disingenuousness, Murdo William McRae has paid close attention to the paradoxes Levi thus inscribes into his text and has developed what might paradoxically be called an affirmative deconstructive reading, along the following lines:

> All of Levi's disclaimers about the autobiographical status of his text are motivated by the fundamentally non-narrative structure of his meditations. If Levi's text records something like the temporal progress of his life, that record is not dominantly structured by narrativity. Levi's meditations record instead a life lived less in narrative time than in textual space, where the boundaries are marked by the

periodic and cyclic inscription, reinscription, and reversal—the deconstructive play—of the oppositions of matter and spirit on the one hand and language and reality on the other. (McRae 1988, 116)

McRae's remarks go directly to the intellectual center of *The Periodic Table*, what he terms a problematic "Platonic strand" (118) that one can clearly trace throughout the text. In this essay I will build on these observations. To begin with, although Plato is famously credited with initiating canonical forms of Western logocentric discourse and thus the dominant philosophical moralizations of the polarities between spirit and matter, and reality and language, there is also a relatively muted Platonic discussion that posits a third and mediating term, and that thus disrupts the strictly dualistic structures of logocentrism from within. Into the ontotheological dualisms of transcendent/mundane and divine/human, Plato occasionally inserts the mediating category of the daemonic.[3] I will attempt to tease out of *The Periodic Table* its implicit and explicit investments in the structures and functions of the daemonic. In the process I will supplement McRae's deconstructive recuperation of Levi's "oppositions and reversals" with a reading that returns the text to both its chemical and its autobiographical narratives.

By *daemonic* I am stipulating neither a spiritual nor a moral process but a certain alchemy of relationships, a dynamic of structures.[4] In this view, the daemonic is not the satanic shadow and corruption of divine origins, but rather that narrative register concerned with the indeterminacy and changeability of forms. Described not as an experience but as a textual construct, the mythopoetic realm of the daemonic is a structural allegory for systems of mobility, mediation, and transformation; it is an extended anthropomorphic figure for networks of relation and communication. Consider Diotima's description to Socrates of Eros as a daemon in Plato's *Symposium*:

"He is a great spirit (*daimon*), and like all spirits he is intermediate between the divine and the mortal." "And what," I said, "is his power?" "He interprets," she replied, "between gods and men, conveying and taking across to the gods the prayers and sacrifices of men, and to men the commands and replies of the gods; he is the mediator who spans the chasm which divides them, and therefore in him all is bound together." (§202e; Plato 1956, 44)[5]

Whether angel or devil, as a figure of communication the daemon is a per-sonified message moving rapidly through space and across borders: thus it is depicted as winged.

In keeping with winged Eros as a prime figure of the daemonic, I will discuss some of the "erotic" elements in *The Periodic Table*, for it seems clear to me that Levi downplayed the autobiographical component of this text to some extent to deflect attention from and mute the memoir of sexual adolescence embedded within it. Although the figure of Eros per se never makes an explicit appearance in the text, another daemonic figure from pagan mythology is in great evidence throughout and will serve as the touchstone for my reading. It is Mercury, a god who gives his name to a heavy metallic element, the peculiarly fluid quicksilver to which a signifi-cant chapter of *The Periodic Table* is devoted. Briefly, Mercury, the Roman version of the Greek god Hermes, is Zeus/Jove's messenger and the patron of trade, science, rhetoric, and theft. Hermes/Mercury is also the forebear of Hermes Trismegistus, whom Levi invokes several times as the patron of modern chemistry's disreputable alchemical beginnings.

Although initially and nominally Hermes/Mercury is an Olympian divinity, nevertheless, in his roles as messenger, delegated herald, mediator, purveyor of commodities, fast talker, inventor, and metamorphic trans-former, he is essentially a daemonic figure. His particular identifications with commerce, theft, eloquence, and science epitomize the daemonic func-tion in its various aspects, as well as offering a decent outline of the main themes addressed in *The Periodic Table*. We need only add the element of the erotic, which drops somewhat away from Mercury but is bound up with Hermes as lover of Aphrodite—to the point of virtual fusion in the figure of Hermaphroditus. What must be emphasized about the daemonic function is its ambivalent intermediacy. It tends to come between things and complicate as well as facilitate their relations, as sexual desire may arise between two persons for whom it is illicit, or as commerce comes between commodities and their consumers, thieves between persons and their possessions, rhetoric between meanings and their communication, and science between divine mystery and human ignorance.

One other daemonic scenario needs mention here, the one constellated by Hermes/Mercury's relation to Zeus/Jove. The relation between sons and fathers, or, more generally, between children and parents, is allegorized by the relationship between the daemonic and the divine. The son/child is an ambivalent supplement in a secondary or minority relation to patri-archal/parental primacy and majority. In the adolescent child, the shifty intermediateness of the filial situation is exacerbated. However, as myths of Hermes' behavior indicate, although the divine father may issue his commands, the daemonic son is at liberty to disobey and lie about it later.[6]

To return to *The Periodic Table*, it seems clear on the face of it that Levi is familiar with some of these mythological allegories of the daemonic. For instance, consider the point in the text when the figure of Mercury is first explicitly mentioned, in the chapter "Nickel," which tells of the time in late 1941 when as a twenty-two-year-old college graduate, Levi took a job devising a method to extract nickel from the discarded rocks of an asbestos mine:[7]

> The doorbell rang—it was a tall, thin young man wearing the uniform of the Italian army, and I immediately recognized in him the figure of the messenger, the Mercury who guides souls, or, if one wishes, the annunciatory angel. In short, the person for whom everyone waits, whether he knows it or not, and who brings the heavenly message that changes your life for good or ill, you don't know which until he opens his mouth. . . . Who had sent him to me? Another Mercury, Caselli, the inflexible custodian of another man's fame. (62)

What Levi underscores here is that the immediate messenger delegated to bring him the job offer had been sent by another delegate, "another Mercury, Caselli." At this point a backward light is directed on an oddity in the earlier chapter "Zinc," which recounts Levi's first year at the Chemical Institute as a student of "Professor P." There Levi gives a detailed analytical description of P's "faithful technician-beadle" (30), Caselli, and of the attachment that had taken hold between the chief scientist and his technical functionary: "Caselli loved P. with a bitter, polemical love. Apparently he had been faithful to him for forty years; he was his shadow, his earthbound incarnation, and, like all those who perform vicarial functions, he was an interesting specimen: like those, I mean to say, who represent Authority without possessing any of their own" (32). Although the erotic overtones here are lightly ironic, they also point to the daemonic complexion of the specimen being examined—the role Caselli enjoys as herald and "earthbound" signifier for the implicitly "divine" authority of the great professor.

My speculation is that Levi's recognition of Caselli as "another Mercury" is a kind of retroactive interpretation or *après coup*, what psychoanalysis terms *nachträglichkeit*, or deferred action, enabling Levi to reinvest a heightened interest in the phenomenon of Caselli and expressing on Levi's part a certain degree of identification with this exemplar of a dae-monic function. Having broached psychoanalytical topics, I will add that the daemonic as I have been treating it is obviously related to the genera-tional intrigues of the Oedipus complex. But I contend that the daemonic subsumes the Oedipal, and not the other way around. That is, as I now pro-ceed to tease out of Levi's memoirs a certain Oedipal configuration, I do so not to reduce his case to the hackneyed parameters of a vulgar Freudi-anism but to add that family-romantic component to the more complexly significant matter of Levi's investments in the daemonic.

Let us note, then, that the many affectionate vignettes of relatives in the opening chapter, "Argon," are devoted almost entirely to removed gen-erations. Only at the very end of the chapter does Levi mention his own parents. The two short scenes involving his father are both symbolically charged. In the first, the boy remembers his father as a habitual violator of kosher restrictions: "he liked prosciutto so much that, faced by the temp-tation of a shop window, he yielded every time, sighing, cursing under his breath, and watching me out of the corner of his eye, as if he feared my judgment or hoped for my complicity" (19). Did the young Levi indulge or withhold that complicity with his father's guilty acts? In any event, the dynamics of this scene are curiously inverted, in that the son's memory positions him as standing in judgment of the father's petty transgressions. The second scene with Levi's father immediately follows and also contains the sole direct mention of his mother: "One hardly ever saw [Grandmother Malia's husband] the doctor, nor did I certainly want to see him, ever since the day on which I had surprised my father telling my mother that, when they brought him stammering children to be treated, he would cut the fillet of skin under the tongue with his scissors" (19–20). The Oedipal milieu here is invoked by the striking overlapping of a primal scenario—the child "surprising" his father in (verbal) intercourse with his mother—with a castration emblem: the conversational scissors recalled in aggressive and wounding proximity to an indispensable (verbal) organ. In both cases, what distorts the Oedipal configuration is some metonymy of speech, as the secondary linguistic process overlays the primary erotic processes.

I dwell on these details in order to frame the rhetoric Levi uses to de-scribe his relations with chemistry and evoke the romance of his intellec-tual life. The polar oppositions on which McRae has commented and by which Levi meditated about the significance of his pursuits are easily analo-gized to gendered familial roles, at times by Levi himself. The terms of this

romantic agon are enunciated as follows: "Caselli handed me my zinc. . . .
The moment, desired and somewhat feared, had come. The hour of the
appointment with Matter, the Spirit's great antagonist, had struck" (33).
The pursuit of an absolute order or Spirit is the patriarchal commandment
clearly inscribed when Levi remarks how "chemistry represented an indefi-
nite cloud of future potentialities which enveloped my life to come in black
volutes torn by fiery flashes, like those which had hidden Mount Sinai. Like
Moses, from that cloud I expected my law, the principle of order in me,
around me, and in the world" (23). Matter assumes the corresponding place
of an Oedipally disbarred but perennially attractive maternal femininity, a
place explicitly named a bit later when Levi states that in his second year
at the Chemical Institute,

> the affair had turned serious, the confrontation with Mother-Matter,
> our hostile mother, was tougher and closer. . . . Here the relationship
> with Matter changed, became dialectical: it was fencing, a face-to-face
> match. Two unequal opponents: on one side, putting the questions,
> the unfledged, unarmed chemist, at his elbow the textbook by Auten-
> rieth . . . on the other side, responding with enigmas, stood Matter,
> with her sly passivity, ancient as the All and portentiously rich in
> deceptions, as solemn and subtle as the Sphinx. (38–39)

The romance Levi constructs here rewrites great Western epic narratives of heroic masculine confrontation with a feminized unknown. A prior
allusion to *The Odyssey*, a mention of Proteus, had already keyed in this
particular epic register. Speaking for himself and Enrico, his high school
companion, the youthful Levi dreams: "We would be chemists, Enrico
and I. We would dredge the bowels of the mystery with our strength, our
talent: we would grab Proteus by the throat, cut short his inconclusive
metamorphoses from Plato to Augustine, from Augustine to Thomas, from
Thomas to Hegel, from Hegel to Croce. We would force him to speak" (23).
Levi presents the metamorphic daemon Proteus here as a figure feminizing
the textual tradition of philosophical idealism as shifty and "inconclusive,"
over which, through chemistry, Levi as Menelaus would exert heroic scientific control. In the present scenario, Levi/Oedipus has identified again
with the Father/hero's active primacy, casting Mother-Matter into the daemonic position of passive monstrosity. But as an Oedipal signature, the
Sphinx also marks the quest(ion)er as in fact a son figure: our hero is
not an Odysseus yet but still an aspiring Telemachus. The late-adolescent,
Italian-Jewish male chemist-ephebe has uncritically absorbed the hoariest

of Western patriarchal plot lines. He practices articulating them to Sandro, his fellow student at the Chemical Institute:

> Sandro was surprised when I tried to explain to him some of the ideas that at the time I was confusedly cultivating. That the nobility of Man, acquired in a hundred centuries of trial and error, lay in making himself the conqueror of matter, and that I had enrolled in chemistry because I wanted to remain faithful to this nobility. . . . That if one looked for the bridge, the missing link, between the world of words and the world of things, one did not have to look far: it was there, in our Autenrieth, in our smoke-filled labs, and in our future trade. . . . Sandro listened to me with ironical attentiveness, always ready to deflate me with a couple of civil and terse words when I trespassed into rhetoric. (41–42)

It is clear, of course, that with his first step forward into these meditations Levi had already "trespassed into rhetoric," in his uncritical deployments of Hebraic and Greek patriarchal discourses, including the Platonic rhetoric of the demotion of "rhetoric" as a deceptive and fraudulent (daemonically eloquent) "world of words" away from which one must flee toward a real and unchanging (divinely eternal) "world of things." What saves *The Periodic Table* from the mere repetition of these masculinisms are the simultaneous if halting countermovements by which Levi questions these constructions. Although Levi identifies the Italian fascism he loathes with the corruptions of rhetoric, fascism also appropriates Levi's language of Spirit, causing him later to rethink his attitudes toward Matter, and chemistry as its conqueror.[8] At the level of autobiographical development, the undergraduate Levi was groping his way toward a recognition that his chosen world was equally the world of words as well as the world of things, that as an intellectual being he was also a writer-rhetorician, and that his proper function would not be to elevate one world over another but to find "the bridge, the missing link" between them. In the terminology I have been using, Levi was moving toward an understanding of his own ambivalent identification with the intermediating functions of the daemonic.

At this point in the book, however, Levi's dreams of scientific conquest are punctured most concretely by his virtual virginity. His as-yet-untranscended sonhood is continually brought home by the annoying nonentity of his erotic life. The text intimates that the fascist racial laws were not only an affront to Italian society and civilized humanity but also a detriment to Primo's ability to get dates. In "Zinc" he describes his crush on a lab colleague named Rita, with whom he once had the good fortune to walk home. In "Nickel," after he takes the job at the asbestos mine, he begins to learn about the past orgiastic shenanigans of its employees: "many, many stories were told; from what could be gathered, all fifty of the mine's inhabitants had reacted on each other, two by two, as in combinatorial analysis" (68). As for Primo, "I fell in love with my work from the very first day" (71). In the lab he meets Alida: "She too, like all the people up there, had interacted with several persons and did not make a mystery of it with me" (71–72). "Interacted," like "reacted" in the prior quote, is, of course, a chaste locution, a coy chemical euphemism for "had sex." In the midst of the sexual license of the isolated asbestos mine, then, the still-virginal Primo begins his true literary career by penning stories of chemical marriage quests—what he terms "two stories of islands and freedom" (73)—which survived decades of disappearance to be interpolated into *The Periodic Table* as the chapters "Lead" and "Mercury."[9]

These two exercises in the personification of elemental characters bring together the daemonic themes of eros, science, and commerce—intercourse, transformation, and exchange—under the notion of "trade." This sexual/scientific commerce is, I think, the deeper significance of Levi's eventual insistence that *The Periodic Table* is to be considered the "microhistory . . . of a trade." "Lead," the lesser of these two tales, concerns Rodmund, a lead prospector seeking to preserve his line: "*So, after six generations in one place, I began traveling again, in search of rock to smelt or to be smelted by other people; teaching them the art in exchange for gold. We Rodmunds are wizards, that's what we are: we change lead into gold. . . . For us, women serve to provide a male child, so that the race does not die out, but we don't take them along*" (81). Along the way Rodmund meditates on lead as the ashy remains of a long period of (radioactive) transmutation: "*Lead is actually the metal of death . . . a metal which you feel is tired, perhaps tired of transforming itself and that does not want to transform itself any longer*" (87).[10] Similarly, the goal of Rodmund's long journey is to couple and reproduce himself before succumbing to the lead poisoning that has already reached his hands and gums. His quest ends rather bluntly on a rocky island: "*I can't say how, but right there was the lead: I felt it under my feet, turbid, poisonous, and heavy, stretching for two miles along a brook in a wood where wild bees nest in the lightning-*

*struck tree trunks. In a short time I had bought slaves who dug for me,
and as soon as I had laid aside a bit of money I also bought myself a
woman"* (95).

"Mercury" also concerns the purchase of wives through the trade of
elemental metals. However, whereas the first-person narrator of "Lead"
went in search of a wife and found her on an island, the first-person narrator of "Mercury" possesses a wife already and begins his tale on the
island Desolation. Behind this fiction, too, lies *The Odyssey*, the opening
of book 5, where Zeus delegates Hermes to rescue Odysseus from cap-
tivity on Calypso's island. In "Nickel," Levi mentions in passing that his
model for this imaginary island was the forlorn British possession of Tris-
tan da Cunha in the southern Atlantic, but the *tristesse* of this desolate
isolation would also seem to be a figure for the virgin captive/author in the
midst of his eroticized surroundings at the geographically isolated asbestos mine. On Desolation Island, a married couple, Corporal Daniel K. and
Maggie Abrahams, have lived for fourteen years when their solitude is broken by the arrival of two Dutchmen and two Italians. It emerges that the
exiled renegade Hendrik is an alchemist on the lam: *"He and Maggie took
long walks together, and I heard them talk about the seven keys, Hermes
Trismegistus, the union of contraries, and other obscure matters"* (100).

Hendrik personifies Hermes/Mercury in several ways, ranging from the
metaphysical to the commercial realms: ostensibly he aspires to transform
baser matter into gold, but virtually he is an eloquent trickster and a potential thief whom Corporal Abrahams suspects of intending to steal his wife;
in addition, like the *Moly* Hermes delivers to Odysseus in book 10 of
the *Odyssey*, Hendrik will bring Abrahams a *pharmakon* in the form of
mercury. When Abrahams tries to enlist the captain of a whaling vessel,
which puts into Desolation Island once a year, to deliver wives for the four
single castaways, *"he laughed in our faces"* (101). What do they intend
to pay with, he asks, pork sausages? On cue, the island's volcanic mountain promptly erupts, opening up a fissure in which a pool full of mercury
begins to collect:

*Hendrik seemed transfigured. He exchanged swift glances with Mag-
gie whose significance I could not catch, and he said some obscure,
mixed-up things to us, which, however, she seemed to understand:
that it was time to initiate the Great Work; that, like the sky, the earth
too has its dew; that the cave was full of the spiritus mundi. Then he*

turned openly to Maggie and said to her: "Come here this evening; we will make the beast with two backs." (103)

When Abrahams threatens to knife Hendrik over that declaration, the alchemist hastens to explain that the daemonic/erotic figure of the "beast" is an alchemical metaphor for the sexual coupling not of persons but of elements: "*As for the beast, he said it was not something that could be explained in a few words. Mercury, for their work, would be indispensable, because it is a fixed volatile spirit, that is, the female principle, and combined with sulfur, which is hot male earth, permits you to obtain the philosophic Egg, which is precisely the Beast with Two Backs, for in it are united and commingled male and female*" (104–5).

When Hendrik divulges to Abrahams that mercury is also a precious commodity in demand for less esoteric applications, the denouement quickly follows, and Hendrik fulfills his Hermetic function as Abrahams's rescuer and good angel. Supplied with a load of mercury, the whaling captain procures four women and delivers them to Desolation. In the final twist, Maggie/Calypso sets up housekeeping with Hendrik/Hermes, for Abrahams/Odysseus has now found his new Athena: "*The girl with the grey eyes did not displease me, even if she was much younger than I; on the contrary, she made me feel gay and lighthearted, like a tickle, and brought to mind the idea of catching her on the wing like a butterfly*" (108). Enacting a perennial patriarchal fantasy that gathers up the erotic and commercial elements of the daemonic Hermes, here the notion of a "trade" materializes as a wife-swapping in which the narrator-hero outbargains the figure of Mercury.

However, as the next chapter, "Phosphorus," goes on to relate, a year later the author of these daemonic daydreams has fallen in love once again with a female lab colleague, Giulia Vineis, who is willing to champion his cause in the role of go-between: "And what about me? No? No girls? That's bad: she would try to help me out there, forget the racial laws; a lot of nonsense anyway, what importance could they have?" (113). The signature of Hermes as daemonic intermediary has been passed on to Giulia, for "this Giulia was a bit of a witch—she read palms, went to mediums, and had premonitory dreams—and sometimes I dared to think that this haste of hers to free me of an old anguish and procure for me immediately a modest portion of joy came from a dark intuition of hers about what fate had in store for me, and was unconsciously aimed at deflecting it" (117–

18). In other words, Levi was still a virgin at twenty-three. One evening, having transported her on his bicycle handlebars to the apartment of her fiancé, and while waiting on the street in despair and mortification for her return: "I fell back on what was my dominant thought during those years: that the existing fiancé and the laws of racial separation were only stupid alibis, and that my inability to approach a woman was a condemnation without appeal which would accompany me to my death, confining me to a life poisoned by abstract, sterile, and aimless desires" (125).

As matters worked out, despite Giulia's efforts Levi was caught up in World War II before that situation altered. In "Chromium," several chapters later, Levi has survived Auschwitz and returned to Turin. In a distinctly anticlimactic manner, the comment wedged into an anecdote about cans of livered paint, Levi briefly narrates the climax of his own protracted marriage quest: "Now it happened that the next day destiny reserved for me a different and unique gift: the encounter with a woman, young and made of flesh and blood. . . . In a few hours we knew that we belonged to each other. . . . In a few hours I felt reborn and replete with new powers, washed clean and cured of a long sickness. . . . My very writing became a different adventure" (153). Here, then, long after the duress of horrendous international upheavals had diverted his attention from his amorous tribulations, is the muted climax of the erotic narrative embedded in Levi's tales of the elements and a certain resolution of his residual Oedipal situation. Still, Levi's intellectual and professional romance with chemistry appears to have derived its emotional intensity to a great degree from the Oedipal arrangements and erotic deprivations of his adolescence. For one last instance, in "Phosphorus" Levi remarks that, "after Giulia's marriage, . . . I felt a widower and an orphan and fantasized about writing the saga of an atom of carbon, to make the people understand the solemn poetry, known only to chemists, of chlorophyll photosynthesis: and in fact I did eventually write it, but many years later, and it is the story with which this book concludes" (128).

"It is exactly to this carbon that I have an old debt, contracted during what for me were decisive days. To carbon, the element of life, my first literary dream was turned" (225). Not surprisingly, the concluding story Levi tells here about the poetry of photosynthesis is an elemental parallel to the cyclical narratives of epic quest romance. In the Homeric/Ovidian tradition in which Levi is grounded, photosynthesis evokes the figures of Hyperion and Phoebus Apollo, the Titanic and Olympian solar engines of all earthly transformations. Putting Levi's psychosexual motivations aside at this point, a question remains: Has he merely adapted the world to his text, or is the structural homology between the wanderings of carbon and the wanderings of the hero in quest of romance an actual insight into the real?

In terms of the daemonic complexion suffusing Levi's text, "Carbon" is a metamorphic romance concerning the wanderings of a most transformative element. If water, the medium of all biological processes, is the great transformative molecule, carbon is the great allotropic atom for organic life. It lends the same atomic structure to chemical forms as varied as graphite and diamond—and combined with water, it issues into living tissue. Carbon "is the only element that can bind itself in long stable chains without a great expense of energy. . . . Therefore carbon is the key element of living substance: but its promotion, its entry into the living world, is not easy and must follow an obligatory, intricate path" (226–27).

The coincidence of narrative romance with the description of the behavior of carbon begins with the ability of carbon to "promote" itself from inert to living matter, to "insert" itself into living processes. By so doing the fate of the carbon atom literally parallels the fictional or "figurative" promotions—metaphors, personifications, daemonic interventions—whereby a narrator, ingenuously or not, fictively animates the things of the world. As the world of romance arises out of the anthropomorphic promotion of inert terms into animated figures, so life arises with the promotion of carbon from a chemical compound to a biological participant. Thus Levi begins to describe the existence of his chosen carbon atom during its primal sleep—"bound to three atoms of oxygen and one of calcium, in the form of limestone" (225)—with a personification: "Its existence . . . is a pitiless alternation of hots and colds . . . an imprisonment, for this potentially living personage, worthy of the Catholic Hell" (226).

By taking the carbon atom as part of a "potentially living personage," Levi endows that elemental identity with "soul," but carbon's allotropic possibilities and metamorphic nature already make it apt for such figurative promotion. And in terms of this daemonic personification, Levi can elaborate a secular myth of incarnation: the story of the fortunate fall of an elemental life principle. The first sublime crossing of Levi's tale is the leap taken by the carbon atom over the limen from physics to biochemistry, from matter to life. Winged like Mercury, as carbon dioxide "it travelled with the wind . . . now high, now low, on the sea and among the clouds . . . ; then it stumbled into capture and the organic adventure" (226). Now is the moment of blessed transgression or transcendental nomination: the carbon atom "had the good fortune to brush against a leaf, penetrate it, and be nailed there by a ray of the sun" (227): "Our atom of carbon enters the leaf, colliding with other innumerable (but here useless) molecules of nitrogen and oxygen. It adheres to a large and complicated molecule that activates it, and simultaneously receives the decisive message from the sky, in the flashing form of a packet of solar light" (227–28).

Levi has translated mythic intuition by way of scientific vision: Hermes appears once more, mercurial messenger from heaven, carrying the celestial packets, photons of solar energy, into a daemonic world. "Now our atom is inserted: it is part of a structure . . . a beautiful ring-shaped structure, an almost regular hexagon" (228), a molecule of glucose ($C_6H_{12}O_6$), in which carbon and water combine in the joint effort of organic life. But as in a Neoplatonic allegory the immortal soul is only temporarily captured by the flesh, the carbon atom's organic residence is also only a temporary, magical, yet natural accident.

Following out this saga, Levi traces a metamorphic series of forms worthy of the transformations of Proteus: the "organic adventure" of this carbon atom, from leaf to grape to wine, from mouth to stomach to liver to muscle fiber, there to be overtaken by oxidization:

> So a new molecule of carbon dioxide returned to the atmosphere, and a parcel of the energy that the sun had handed to the vine-shoot passed from the state of chemical energy to that of mechanical energy. . . . "Such is life," although rarely is it described in this manner: an inserting itself, a drawing off to its advantage, a parasitizing of the downward course of energy, from its noble solar form to the degraded one of low-temperature heat. In this downward course, which leads to equilibrium and thus death, life draws a bend and nests in it. (230)

The vital thermodynamic romance of the carbon atom begins only when it enters "the narrow door of photosynthesis . . . not only the sole path by which carbon becomes living matter, but also the sole path by which the sun's energy becomes chemically usable" (231), but ends through any of the broad egresses of metabolic processes, which little deaths, however, only prepare for further cycles, endless adventures. So Levi concludes by narrating a cycle that ends where chemistry ends, at the limen between life and mind, picking up the story with the same atom having now found its way into a glass of milk:

One, the one that concerns us, crosses the intestinal threshold and enters the bloodstream: it migrates, knocks at the door of a nerve cell, enters, and supplants the carbon which was part of it. This cell belongs to a brain, and it is my brain, the brain of the me who is writing; and the cell in question, and within it the atom in question, is in charge of my writing, in a gigantic miniscule game which nobody has yet described. It is that which at this instant, issuing out of a labyrinthine tangle of yeses and nos, makes my hand run along a certain path on the paper, mark it with these volutes that are signs: a double snap, up and down, between two levels of energy, guides this hand of mine to impress on the paper this dot, here, this one. (233)

We see Levi's contemporaneity in the way he brings his speculations down to the scene of writing, ingeniously turning the final period of his *Periodic Table* (written, one hopes, with a trace of graphite from a pencil) into an iconic sign standing for a carbon atom. "These volutes that are signs" trace a spiral path across the handwritten page that maps the recursive paths of carbon atoms in and out of the organic adventure, as well as Levi's own vortex of recursions to the daemonic romances and erotic processes of epic narratives: "I could recount an endless number of stories about carbon atoms . . . others still that descended to become part of the mysterious shape-messengers of the human seed, and participated in the subtle process of division, duplication, and fusion from which each of us is born" (232). In procreative biochemistry, it seems, Levi grasps the fusion of his own masculinity and femininity, Hermetically resolves his own struggle between the austerity of physical science and the labors of writing. Carbon atoms form genetic "shape-messengers" that fuse Matter and Spirit in a third, daemonic term: human life.

NOTES

1. Unless indicated otherwise, all page numbers in the text refer to *The Periodic Table*. Levi's other works in English are listed in the References.

2. On Lewis Thomas, see Clarke (1990). Levi's work most reminiscent of Thomas's essays, similarly journalistic in origin, is *Other People's Trades* (1989).

3. McRae refers his reading of Levi's Platonisms to "Plato's Pharmacy" (Derrida 1981). The intermediate and indeterminate functions of the daemonic are precisely the burden of Derrida's discussions here of the *pharmakon*, a figure by which

Plato analogizes written texts to elixirs or potions, insofar as both are ambivalent supplements with unpredictable effects.

4. My approach to the daemonic developed out of a study of metamorphosis in literature, whereas Michel Serres's "demon" emerged from considerations of thermodynamics and information theory. Concerning systemic noise [*le parasite*] and its exclusion, J. V. Harari and D. F. Bell explain that

> Serres called this included/excluded third man the *demon*. . . . The parasite, like the demon and the third man, is an integral part of the system. By experiencing a perturbation and subsequently integrating it, the system passes from a simple to a more complex state. Thus, by virtue of the power to perturb, the parasite ultimately constitutes, like the *clinamen* and the demon, the *condition of possibility of the system*. In this way the parasite attests from within order the primacy of disorder; it produces by way of disorder a more complex order. (Serres 1982, xxvi–xxvii)

5. Compare W. R. M. Lamb's translation of the same passage to Benjamin Jowett's: according to Diotima, Eros is "a great spirit, Socrates: the whole of the spiritual . . . is between divine and mortal . . . interpreting and transporting human things to the gods and divine things to men; entreaties and sacrifices from below, and ordinances and requitals from above; being midway between, it makes each to supplement the other, so that the whole is combined in one" (Plato 1946, 178).

6. In the *Homeric Hymn to Hermes*, the infant god metamorphoses to pass through a keyhole after stealing Apollo's cattle, then uses his eloquence to move Zeus to wink at his theft; see Boer (1970, 26–61); also Brown (1969).

7. Levi pauses to explain that "the entrails of the earth swarm with gnomes, kobolds (cobalt!), *nickel*, German 'little demon' or 'sprite' " (64).

8. McRae (1988) gives close attention to this turn of affairs in the text: "As the Fascist grip tightened even more on Italian Jews, chemistry came no longer to be a source of certainty in an uncertain world: 'It led to the heart of Matter, and Matter was our ally precisely because the Spirit, dear to Fascism, was our enemy' " (118).

9. Both "Lead" and "Mercury" are italicized in the text of *The Periodic Table*, presumably to distinguish them as found pieces of early writing from the surrounding text of later reflections.

10. Freud also places lead in a series of death symbols: "In Bassanio's short speech [in the *Merchant of Venice*] while he is choosing the casket, he says of lead . . . : 'Thy paleness moves me more than eloquence.' That is to say: 'Thy plainness moves me more than the blatant nature of the other two.' Gold and silver are 'loud'; lead is dumb. . . . In dreams dumbness is a common representation of death" (1958, 294–95).

REFERENCES

Boer, B. C., trans. 1970. *The Homeric Hymns*. Chicago: Swallow.

Brown, N. O. 1969. *Hermes the Thief*. New York: Vintage Books.

Clarke, B. 1990. Life, Language, and Identity: Lewis Thomas's Biomythology in *The Lives of a Cell*. In *The Body and the Text: Comparative Essays in Literature and Medicine*. Ed. B. Clarke and W. Aycock. Lubbock: Texas Tech Univ. Press.

Derrida, J. 1981. Plato's Pharmacy. In *Dissemination*. Trans. B. Johnson. Chicago: Univ. of Chicago Press.

Freud, S. 1958. The Theme of the Three Caskets. In vol. 12 of *The Standard Edition of the Complete Psychological Works of Sigmund Freud*. Trans. J. Strachey et al. London: Hogarth.

Levi, P. 1984. *The Periodic Table*. Trans. R. Rosenthal. New York: Schocken Books.

———. 1985. *If Not Now, When?* Trans. W. Weaver. New York: Summit.

———. 1986a. *Moments of Reprieve*. Trans. R. Feldman. New York: Summit.

———. 1986b. *The Monkey's Wrench*. Trans. W. Weaver. New York: Summit

———. 1986c. *Survival in Auschwitz and the Reawakening: Two Memoirs*. Trans. S. Woolf. New York: Summit.

———. 1989. *The Drowned and the Saved*. Trans. R. Rosenthal. New York: Vintage Books.

———. 1989. *Other People's Trades*. Trans. R. Rosenthal. New York: Summit.

———. 1990. *The Sixth Day and Other Tales*. Trans. R. Rosenthal. New York: Summit.

McRae, M. W. 1988. Opposition and Reversal in Primo Levi's *The Periodic Table*. *Publications of the Mississippi Philological Association*. 115–24.

Plato. 1946. *Lysis, Symposium, Gorgias*. The Loeb Classical Library, Plato V. Trans. W. R. M. Lamb. Cambridge: Harvard Univ. Press.

———. 1956. *Symposium*. Trans. B. Jowett. Indianapolis: Bobbs-Merrill.

Serres, M. 1982. *Hermes: Literature, Science, Philosophy*. Ed. J. V. Harari and D. F. Bell. Baltimore: Johns Hopkins Univ. Press.

Tagliabue, J. 1987. Primo Levi, Author on Holocaust, Is Found Dead. *New York Times*, 12 April.

Tax, M. 1986. Speak, Memory: Primo Levi's Living History. *Voice Literary Supplement*, March.

West, P. 1984. Formulas for Literary Alchemy. *Washington Post Book World*, 30 December.

Hermeneutics and the New Epic of Science

MARTIN EGER

In a recent paper entitled "Why Is There No Hermeneutics in Natural Science?" the philosopher Gyorgy Markus accepts the widespread belief that "the natural sciences have lost their direct and general cultural significance" and that "the literary objectivations [of these sciences are] not read today by a wider public beyond a narrow circle of professional experts." Reasoning along this line, he concludes that hermeneutics plays no serious role here because today science rests on an institutionalized author-text-reader relation which is normative and impersonally imposed. In short, we now have "cultural closure of scientific discourse upon itself" (Markus 1987, 28). However, by "text" and literary "objectivations" the author means purely professional, technical reports. About popularizations he says nothing, assuming, apparently, that for the serious business of science such literature is peripheral at best.

I do not propose now to blur that well-established distinction between "science itself" and "popular science," between science for peers and science for the general reader.[1] I do intend, however, to challenge this dichotomy as a useful classification in the 1990s, especially in the context of possible roles for hermeneutics. For there exists today at least one other group of scientific writings, a rather influential one, that deserves a name of its own.

Consider that literature which has come to the fore during the past quarter century or so, mostly full-length books combining high-quality writing with scientific depth. Typically it deals with subjects that have philosophical or social or humane implications. It is aimed neither at the specialist nor at the casual reader. It is a body of writing in which scientific and literary seriousness are blended and offered up consciously in an attempt to break through professional barriers with a message deemed important enough to be worth the effort of *demanding* reading. It comes mostly from scientists, especially the famous; and often it involves fairly complicated technical ideas. Of particular importance are its interdisciplinary effect and its impact on serious thinking in nonscientific fields.

My examples are Jacques Monod's *Chance and Necessity* (1971), Manfred Eigen and Ruthild Winkler's *Laws of the Game* (1981), Ilya Prigogine and Isabelle Stengers's *Order out of Chaos* (1984), Steven Weinberg's *The First Three Minutes* (1979), Steven Rose's *The Conscious Brain* (1976), Douglas Hofstadter's *Gödel, Escher, Bach* (1979), Joseph Weizenbaum's *Computer Power and Human Reason* (1976), and E. O. Wilson's *On Human Nature* (1978). Various other books, to which I will also refer, could be included here and grouped according to how well they exemplify a certain ideal type.[2]

Unfortunately, because the scientific/popular dichotomy *is* so potent in our language, such books are still classed as "popularizations." Yet they differ qualitatively from the usual simplified expositions that do not claim to convey a serious message; and they differ also from even the best popular essays like those by Stephen Gould or Lewis Thomas, which, for the most part, are easily digestible and often biographical or personal. In contrast, the works I cite distill the deeper meaning of scientific advances on fairly broad fronts, calling attention to cognitive implications that bear on human self-understanding. In the 1950s a work of this type stood out conspicuously; today it is part of a genre. And so the question arises: If (contra Markus and others) this literature still has, or has *again*, the capacity to mediate between a knowledge of nature and genuine human interests, *then what interests or which modes of life does it address? And what, if anything, does this have to do with hermeneutics?*

Let me begin to answer these questions by reviewing two types of responses offered only a short time ago: the first presented in typically pithy fashion by the physicist Richard Feynman; the second, not so concise, by Jurgen Habermas. Both agree in one important respect: the *nature* of the subject matter of modern science makes it extremely difficult, if not impossible, to relate this matter to the concerns of human life.

In hindsight, we can now say that both men were wrong. And it should be enlightening to try to understand not only what they missed but why they missed it, for that may indicate where the literature of science is heading today and what role hermeneutics might play in it. With this purpose in mind, let us now consider what I will call "Feynman's question," "Huxley's reply," "Habermas's error," and "Wilson's insight." We begin with Richard Feynman.

Poets say science takes away from the beauty of the stars—mere globs of gas atoms. Nothing is "mere." . . . For more marvelous is the truth than any artist of the past imagined! Why do the poets not speak of it? What men are poets who can speak of Jupiter if he were like a man, but if he is an immense spinning sphere of methane and ammonia must be silent? (Feynman 1963, 3–6)

This, incidentally, from a textbook on physics. Of course we all know the conventional reply, which comes in a variety of styles. Aldous Huxley, for example, puts it as follows:

The world with which literature deals is the world in which human beings are born and live . . . in which they love and hate. . . . [The scientist] is the inhabitant of . . . the world of quantified regularities. . . . Until some great artist comes along and tells us what to do, we shall not know how the muddled words of the tribe and the too precise words of the textbooks should be poetically purified, so as to make them capable of harmonizing our *private and unsharable experiences* with the scientific hypotheses in terms of which they are explained. (A. Huxley 1963, 8–9, 107; emphasis added)

Note that both Huxley's reply and Feynman's question contain certain assumptions about the *possible* relations of science to literature and to life. For Feynman, that link must be aesthetic; poetry being the ideal language, he does not see why the forms of expression used in classical mythology cannot also be applied to a scientific description of the world. He hungers, it seems, for a modern Lucretius. Huxley agrees about the language problem but doubts the solution is so simple, because the dissonance between *private experiences* and a symbolic, *abstract* world is too great for the kind of harmony literature tries to bring about.

It has been noticed, however, that from a historical and philosophical perspective, other relations between life and science are possible. In a series of well-known essays, Jurgen Habermas explored one perennially recognized alternative, only, alas, to reject it. According to Habermas, there are just two possible "channels" through which science can bear on "the practical consciousness of a social life-world": by means of technical exploitation and by means of *mimesis* (the Greek conviction that "through the soul's likening itself to the ordered motions of the cosmos, theory enters the conduct of life"; norms of individual human behavior are ascertained first by understanding cosmic proportions). The philosophically (scientifically) educated person would then possess "practical" knowledge, or what Habermas calls "action-orientation" within a human community. And this "channel," this ancient practice of scientific study for the sake of self-formation, was still thought to be open in the nineteenth century—at least in German universities (Habermas 1970, 53).

Today, however, because scientific knowledge is overwhelmingly instrumental, because theory can be effective *without* first having to transform the person *as a person*, that channel is forever closed, says Habermas. In our time, "information provided by the strictly empirical sciences can be incorporated in the social life-world only through its technical utilization." This means that action-orienting self-understanding is now available from historical-hermeneutic study— of the humanities and certain social sciences. But "taken for itself, knowledge of atomic physics remains without consequence for the interpretation of our life-world, and to this extent the cleavage between the two cultures is inevitable" (Habermas 1970, 52). Here, then, is the more analytical answer to Feynman's question, shared by Markus (1987, 45) and widely accepted.

Yet in all this, something must have been overlooked, unless the literature to which I refer—that third category—is far less successful than we have reason to believe. To shed some light on what Huxley missed, what Habermas left out, and why Markus still ignores a phenomenon that is far from marginal, let us look at this literature more closely. Although the books I listed are well known and hardly need additional reviewing, I propose now to highlight briefly certain of their common features, to show why the mood, the attitudes, and the message of these scientist-writers is (for the most part) quite different from what one might have gathered by reading Huxley and Habermas in the past, or Gyorgy Markus now.

The first similarity is a certain overall structure typical of the most impressive works of this genre. Not that structure will be our main consideration. The argument will turn more on content than on form. But the link between these two aspects of a text is too well known for the latter to be ignored even when the former is central. Moreover, I want to call attention to the different *kinds* of resonance that exist between the works we are considering; not only the message but a certain rhythm in proclaiming that message is shared. Let me call that rhythm "P-S-P," for philosophy-science-philosophy. In its pure form, the book begins with a philosophical survey, sometimes only an introduction, in which certain human or social problems are posed and related to some aspect of science; that science is then expounded in a lengthy technical or quasi-technical section, and the work ends with the original philosophical problems to which solutions or responses are now offered *on the basis of the scientific content just presented*.

The initial discussion, the first "P," often refers to the two-cultures split explicitly, promising new ideas that might heal the wound. For example, "In the blending of biology and the social sciences . . . the two cultures of Western intellectual life will be joined at last" (Wilson 1978, 10); or: "The alienation of science from life . . . is about to be overcome by the evolution of science itself" (Yantsch 1981, 5). These philosophical introductions usually pose other kinds of dichotomies as well, such as objectivity versus animism (Monod), scientific materialism versus outdated mythologies (Wilson), timelessness versus time-consciousness (Prigogine and Stengers), chance versus necessity (Monod, Eigen), or formal logic versus human reason (Weizenbaum). In some of the works, a conceptual contrast of the latter kind may be linked directly to the two-culture split. Thus, Prigogine and Stengers say quite bluntly that "the dichotomy between the 'two cultures' is to a large extent due to the conflict between the atemporal view of (classical) natural science, and the time-oriented view that prevails in a large part of the social sciences" (1984, xxviii). As a rule, both the theme and the underlying mood of the authors are very serious, even when (as in the case of Hofstadter or Eigen and Winkler) a lighthearted mode of exposition is used. The aim is not just enlightenment but something closer to *revelation*.

The meat of the P-S-P sandwich, the scientific middle portion, ranges over all the disciplines, from flip-flops to Church's thesis, from allosteric reactions to space-time singularities, and much more. Without a doubt, resolute attempts are often made in these books to convey some scientific knowledge, considered valuable in itself and needed to develop the philosophical points. Yet the most striking feature is the flagrant excitement: the repeated assertions that only within the *recent* past have new discoveries made it possible to treat such philosophical and social subjects scientifically, and the eagerness of the authors to spread these insights beyond their own specialized community. "The selective theory of evolution did not take on its full significance, precision, and certainty until less than twenty years ago," says Monod (1971, 24). Similarly, Weinberg: "Throughout most of the history of modern physics and astronomy there simply has not existed an adequate observational and theoretical foundation on which to build a history of the early universe. Now, in just the past decade, all this has changed" (1979, 2). In general terms, of course, such developments are widely known, but these authors see similar things happening not just in molecular biology and astronomy but throughout the sciences; and the cumulative effect causes the exhilaration. Thus, Steven Rose: "Only recently have the tools become available to approach the problem of the brain in a way which makes it possible to ask meaningful questions" (1976, 46). And in *Order out of Chaos*, based on *chemical* research, we read once again, "In the past few decades, something very dramatic has been happening in science, something as unexpected as the birth of geometry or the grand vision of the cosmos as expressed in Newton's work. . . . On all levels, from elementary particles to cosmology, randomness and irreversibility play an ever increasing role. *Science is discovering time*" (Prigogine and Stengers 1984, xxviii).

The final, philosophical "bread" of our sandwich includes unabashed *calls for a new morality* or a new "vision" of the world. In this regard, many of the authors view themselves as members of a like-minded group, as contributors to a historically continuous movement in which every text builds on the others, amends the others, and reflects back on common intellectual forerunners.[3] I will refer to these more committed authors as the "core group." Thus, Monod's version of an "ethic of knowledge"—similar to that of George Gaylord Simpson (1949, 339–49)—posits *objectivity* as the first value and foundation of moral "authenticity," from which lower-level goals may be derived. Wilson, of course, goes further, including among his values the diversity of the gene pool, universal human rights (because we are *mammals*!), and the genetic transformation of the human species to forms of higher intelligence and creativity. All this, we are to understand, follows in some serious sense from the science of the middle portion of the sandwich. Without that portion, the tone of authority in this literature would have no visible foundation; and that precisely is why the structure of the texts so often adheres to the format just outlined.

But let us proceed to the message itself, for that is what truly unites the authors of the core group, making of each individual work a contribution to something like a common project. What I shall try to show is that while the literary products of this group derive from widely different sciences, although they come in a variety of styles, and their authors differ on various points, taken together, we have here the telling of *one and the same story*.

It is the story of evolution: evolution explicated in greater detail than ever before, deepened, unified, extended far beyond biology—"universal" or "cosmic" evolution. Although it is an old story, composed gradually with enormous labor, much of it was still not in place even within the last thirty years. Beneath an apparently adequate first-order explanation lurked a number of difficult problems, discerned occasionally even by laymen. It was possible to continue improving various segments of the theory piecemeal, ignoring the problems or shelving them for later, but the result was a partial, patchy picture, not beyond questioning by such luminaries as Niels Bohr, Wolfgang Pauli, Ludwig Wittgenstein, or Karl Popper.[4] The ultimate goal, a truly seamless, thoroughly convincing, all-inclusive science of development—if possible at all—still lay in the future.

Now, however, the excitement evident in the books I am discussing results from a feeling that *this goal is being attained*, that we are living in a time of "the last frontier," of the crucial unification of the scientific worldview, of the dissolution (at least in principle) of the final mysteries. And that feeling is further heightened by what we find in the corresponding literature of the nineteenth century. Even before Darwin, and quite independently among thinkers of all sorts, the idea had arisen that some sort of "development" was going on in line with physical law, encompassing everything from stars to molecules to man. In the writings of Robert Chambers (1845), Herbert Spencer (1880), John Fiske ([1874] 1902), Ernst Haeckel (1905), and others, we see (to our surprise, perhaps) the beginnings of what is now in full swing.[5] It is not just that "cosmic evolution" was a concept much in vogue, but more impressively that (already then!) they "had it right," so to speak, about many specifics of the mature theory as well as its overall shape. That is why today when Eric Chaisson, an author on my larger list, finds kindred thoughts in the work of Robert Chambers, for example, his passion to communicate this discovery can hardly be contained (1987, 94–97). Against the historical background, twentieth-century achievements appear as so many pieces of the jigsaw puzzle falling into place, so many promises kept.

THE THIRD GENRE OF SCIENCE LITERATURE

Among the oldest of mysteries is *purposiveness*. Granted natural selection, how do mindless molecules make up those intricately coordinated, purposeful structures and processes for selection to work on? Monod's *Chance and Necessity* (which sold more than 200,000 copies in French alone) addresses this question by showing that purposiveness (teleonomy) is now unambiguously traced to "microscopic cybernetics," biological feedback mechanisms utilizing certain stereospecific chemical reactions. So crucial is this point for philosophical conclusions, and so interesting, that we find it demonstrated anew, at different levels of organization, by nearly everyone. Douglas Hofstadter, for example, is concerned mostly with other things, but here he is, in *Gödel, Escher, Bach*, making the same argument—for the kingdom of ants!

> *Achilles:* Oh wait. Either the behavior is purposeful or it is NOT. I don't see how you can have it both ways.
>
> *Anteater:* Let me explain. . . . Let's say a signal [team of specialized ants] is moving along. As it goes, the ants which compose it interact . . . with ants of the local neighborhoods. . . . The signal will remain glued together as long as the local needs are different from what it can supply; but if it CAN contribute, it disintegrates, spilling a fresh team of usable ants onto the scene. . . .
>
> *Achilles:* . . . I'm beginning to see. . . . From an ant's-eye point of view, a signal has NO purpose. The typical ant in the signal is just meandering around the colony, in search of nothing in particular, until it finds that it feels like stopping. . . . No planning is required, no looking ahead. . . . But from the COLONY's point of view, the team has just responded to a message. (1979, 320–21)

With teleonomy reduced to the material process of efficient causation, with intermolecular forces as the ultimate *anima*, the appearance of purpose in organisms does not imply the separate existence of some motive power, nor any extraordinary law for living matter. The long history of "animisms," crude or subtle, religious or Marxist, must finally come to an end, says Monod. Only a sober, austere materialism can today do justice to what we have learned. Thus an accomplished biochemist provides scientific grounding for a familiar existential position: In our cold, uncaring universe, man's "destiny is nowhere spelled out, nor are his duties. . . . It is for him to choose" (Monod 1971, 180). Accordingly, what we now ought to choose, he says, is an "ethic of knowledge," for this at least will finally overcome the disastrous split between knowledge and values.

Inescapably, it seems, the strictly biological problem of teleonomy is linked to a larger question of purpose: the meaning, or lack of it, in human life itself. If, to take one example, chance mutations are the ultimate source of the rise of consciousness, then is not this whole story—our being here, our "ascent"—a mere quirk, as Monod thinks, improbable and devoid of purpose? And is not that, in turn, reflected in the fragmentary, jerky quality of individual existence, admittedly part of the nature of things but exacerbated by the character of our civilization and our man-made sciences? The search for some sort of meaningful *unity,* as a major theme in philosophy and culture, is hardly new. What surprises is the powerful reemergence now of this same theme in the *literature of science.*

In 1949 Nobel Prize–winner Hermann Hesse published his monumental novel, *Das Glasperlenspiel* (*The Glass Bead Game*). Although central to the novel, Hesse described this game only vaguely while hinting that its function was to unify different cultural spheres at a deep level. About thirty years later, in a most unexpected way, Manfred Eigen (another Nobel laureate) and Ruthild Winkler gave precise scientific form to their conception of just such a game (Eigen and Winkler 1981). But this construct, involving the same two elements that interest Monod, chance and necessity, is far from fictional. The authors' goal is nothing less than to demonstrate *necessity* in the working of evolution: to remove that existentially disquieting element of contingency, *and,* at the same time, to show that at different levels of organization (including the prebiotic, the linguistic, and the economic) historical processes all have essentially the same deep structure. Beneath apparent chaos, these authors claim, there is a form of unity after all, and of beauty.

The game is one of chance, employing dice, but with the important proviso that its *rules* are chosen for a purpose. In some of the most striking versions, called appropriately "Selection" and "Survival," the rules for placing and removing beads from the board are designed to embody the characteristics of birth, death, mutation, competition, and Darwinian selection. The resulting "play," the increase or decrease of different-colored beads (representing species), becomes then a simulation of the evolutionary process. What gives the idea added glamour is that such games have, of course, been programmed on computers; and the results of many "plays"—suggestive, intriguing patterns—are repeatedly and lavishly displayed, in full color, alongside the ubiquitous Escher.

The scientific message here is subtle: the game concept applies to just about everything. Random events taking place under particular constraints (rules) yield particular types of order, in evolution and in other natural and human processes. "It is still a matter of chance which mutations occur, and in what sequence. But it is a matter of predictable necessity that . . . selection will occur . . . defining a gradient of 'value' . . . for improvement over time" (Eigen and Winkler 1981, 59).

These bead games embody an extraordinary unification. One and the same activity offers satisfaction to scientific curiosity, to the aesthetic sense, and to philosophical interests—a resounding variation on the theme of

Hermann Hesse. Thus, in its wide embrace, its multidimensional approach, *Laws of the Game* represents well a major feature of the genre itself. For despite some disagreement on the role of chance as against necessity, despite differing degrees of enthusiasm for the significance of artificial intelligence (AI), and despite important differences of interpretation (about which more later), the impressive fact is the *convergence* of these authors on a well-defined set of interrelated themes: *extension* of the "evolutionary paradigm" as far as possible, *unification* of the sciences, and *reconciliation* of science with a specifically human reality.

In *Order out of Chaos* we see *all* these themes, developed even more energetically and in full awareness of the goal. First, by tracing out the "tragedy" of how our world progressively split in two—quality versus quantity, sensible things against abstractions—Prigogine and Stengers set the stage in historical-philosophical terms. Next, like directors of a mystery play, they make sure the audience understands what the trouble is and who is hurting. The trouble is the prevailing scientific myth, still the *timeless*, deterministic, Newtonian machine-world. And the pain, which affects all of us, is caused by our inability to find an acceptable human role within the context of such a mechanical worldview. The analysis is familiar, but the recommended strategy is quite new. In his own work and that of others on the same trail, Prigogine sees a *scientific* solution to the *sociomoral problem*. An entirely new paradigm, by restoring *time* to its proper place, will help to heal the old cultural split and erase (again!) that pernicious distinction between scientific and ethical values (Prigogine and Stengers 1984, 312). How does science accomplish such a thing? By attending to the other two themes mentioned above: evolution and unification of our view of nature.

It was the achievement of Prigogine and his collaborators in chemical kinetics to show that even in the macroscopic realm (that is, aside from quantum mechanics), the behavior of a system sometimes reaches what is now known as a "bifurcation point," where a "choice" is made between quite different but equally possible future states. Yet this choice is governed not by law but by small, unpredictable fluctuations. Especially significant for prebiotic evolution is the fact that sometimes these choices lead from disorder to highly organized structures: "order out of chaos," triggered, it seems, for no particular reason that science can discern. Here is that element of chance treated formally by Eigen and Winkler, punctuating the periods of predictable causality. The universe, it turns out, is not a "machine" because it is only *piecewise* determined. As a result, the backward direction of time is not equivalent to the forward direction, even in physics, and the "evolutionary paradigm" (organic, not mechanistic) applies to all the sciences without exception.

In another realm still, that thesis is powerfully supported by Weinberg's much-quoted little book, *The First Three Minutes*. Again we hear of a dramatic confrontation between "two chief world systems," one eternal, stationary, and timeless (steady-state cosmology), the other evolutionary (the big bang). And once more, crucial evidence piles up rapidly in favor of the evolutionary story, now extending back to the first one-hundredth of a second of the universe's existence. As the unification of biology with chemistry gave us in detail the structure of DNA, so now high-energy physics combined with relativistic cosmology yields "the new genesis," not just as ingenious speculation but as a highly worked out scenario. At last, as Einstein used to say, we are getting closer to the "secrets of the Old One." Neither the stars above nor the moral law within are beyond our reach. Once these antipodes symbolized the mysteries of being and pointed to a power beyond. Today, as science accelerates, the assault on both fronts is going rather well.

That this success owes much to the momentum and ingenuity of the computer sciences is widely known. Not so well appreciated, perhaps, is the influence of just the *concepts* and the *metaphors* involved, as distinguished from raw hardware power. It is the merit of books like *Gödel, Escher, Bach* that they convey those concepts in an exceptionally vivid manner and make that influence more understandable. What Kurt Gödel, M. C. Escher, and J. S. Bach had in common, according to Hofstadter, is that each in his own field noticed, appreciated, and utilized brilliantly the concept of self-reference, of a thing turning back on itself, or supporting itself, or reflecting itself, like the scenes in Escher's lithographs.

Again, as with Eigen and Winkler, the aim here is at once scientific, aesthetic, and philosophical: not only to show by means of pictures, stories, riddles, and musical notation how many different and surprising forms this self-reference can take, but to drive home the point that herein lies the deepest of all deep truths ever encountered by the human mind. The "strange loop" is offered to us by Hofstadter as quite possibly the answer to the problem of consciousness.

How does *any* information-processing system answer questions about what it is doing? It sets up within itself a model of itself and then examines that model, examines itself. As Hofstadter puts it, a "*self*-subsystem . . . can play the role of 'soul' . . . in communicating constantly with the rest of the subsystems and symbols in the brain . . . [in] the monitoring of brain activity" (1979, 387–88). Machines do it crudely; animals do it better; in humans, just the complexity of neural hardware is enough for all sorts of unexpected effects.

> Emergent phenomena in our brains—for instance, ideas, hopes, images, analogies, and finally consciousness and free will—are based on . . . an interaction between levels in which the top level reaches back down towards the bottom level and influences it, while at the same time being itself determined by the bottom level. In other words, a self-reinforcing "resonance" between different levels. . . . The self comes into being at the moment it has the power to reflect itself. (1979, 709)

Hofstadter does not quote Kierkegaard here, though he could have done so: "Man is spirit. But what is spirit? Spirit is the self. But what is the self? The self is a relation which relates itself to its own self, or it is that in the relation [which accounts for it] that the self relates itself to its own self; the self is not the relation but consists in the fact that the relation relates itself to its own self" (Kierkegaard 1954, 146).

Now, however, as the philosophical question acquires a scientific answer, computer science and biology converge. Insofar as consciousness is a biological problem, to be accounted for by evolution, the *natural emergence* of such self-modeling "strange loops" out of the growing complexity of organisms looks like a truly saving idea, just what was needed. At last the fog is lifting. The entire panorama is coming into sight, and the "secrets of the Old One"—from big bang to consciousness—seem within our grasp. So rapidly is the picture changing that most of us have not yet realized what is going on.

THE NEW EPIC

How the various contributions fit in as pieces of the big picture is now clear. From Darwin's original theory, the lines of extension radiate downward to prebiotic (chemical) evolution as expounded by Prigogine and Eigen; to cosmic evolution as described by Weinberg, Paul Davies, and the astrophysicists; to human culture as Wilson explains in his theories of sociobiology; and finally, through the work of brain physiologists and AI researchers, to consciousness itself. It is, as Monod would have it, an "objective" story, free of odious dualisms, incorporating "scientific materialism" at its foundation.

Nevertheless, and quite surprisingly, this story too is called a "myth" by the foremost author on my list, Edward O. Wilson. It is Wilson who, seeing an "epic" in the emerging scientific tale, gave explicit articulation to what was sometimes only elliptically said; and it is he who drew conclusions in their most challenging form. This epic, in his view, plays the same role and has much else in common with those very myths it is now displacing. Wilson's vision of the extended, fully worked out, "seamless" theory of cosmic evolution is not normal science—it is the vision of a narrative of unprecedented scope and persuasive power, "far more awesome than the first chapter of Genesis or the Ninevite Epic of Gilgamesh" (Wilson 1978, 209). The evolutionary epic is simply the "best" of myths, according to Wilson, and the one "destined" to prevail because, unlike its predecessors, it does *rest* on science, even if it is not itself science.[6]

In this perspective, the literature I am discussing takes on an obvious and rather serious role. So vast is this new epic, and so detailed, that no one book can encompass it. What we have instead is a large number of major and minor works, on various levels of "scientific literacy," each telling some part of the story, or commenting on it, or interpreting it. Yet taken together, all these constitute the epic itself, which in turn rests on the more solid bedrock of the technical "objectivations" discussed by Markus.

Books that have had a strong impact and have followed closely the ideal type outlined above, both in form and in the orthodox telling of the tale itself, I have called the "core group." But to complete the picture we must keep in mind that numerous others are part of the genre. Eric Chaisson's *Cosmic Dawn* (1981) and *The Life Era* (1987) belong here if taken together, though the scientific and philosophical components are offered more or less separately, each in its own book. Yet Chaisson, an astronomer and leading enthusiast, finds in the new evolutionary epic precisely what the old "animist myths" proffered: "Only when complex organisms arrive at the dawn of the Life Era does *The Universe* acquire self-awareness. . . . 'Matter has reached the point of beginning to know itself.' . . . This, for me, is life's purpose and meaning, its raison d'être—to act as an animated conduit for the Universe's self-reflection" (1987, 229).

Roger Sperry, the famous brain physiologist, has also contributed to the genre. His *Science and Moral Priority* is so concentrated on evolutionary ethics that he nearly skips the technical part altogether (relying, no doubt, on widespread familiarity with his split-brain discoveries). For Sperry, as for Monod, since science now offers a cosmic view of life that "renders most others simplistic by comparison," a new ethics is desperately needed, one that "would lead to the designation . . . of what is good, right, or to be valued morally, as that which is in harmony with, sustains, or enhances the orderly design of evolving nature" (1985, 50). And so on. I skip over many other deserving works. The consensus is overwhelming, perhaps somewhat boring.

However, lest this be the final impression, and the whole genre seem trite, let me end the survey with a dissenting voice: Joe Weizenbaum of MIT, a prophetic figure capable of using the words *God* and *grace* without awkwardness, has made for himself a special place in this literature. The attention received by *Computer Power and Human Reason* (1976), though remarkable, is nonetheless quite understandable in view of its content. Here is a professor of computer science, a member of the leading AI research community, challenging the basic faith of that community head-on—technically, ethically, philosophically. He simply rejects the idea that human reason can be properly *understood* by means of formal logic or computation alone. It is not just that the achievement of current AI programs is *mis*understood and therefore overrated. The more basic problem lies in our failure to distinguish between the computational aspects of thinking and judgment, on the one hand, and, on the other, those aspects which are inseparable from humanness as such. What is human about our reason, says Weizenbaum, is our individual and species *experience*. No other information-processing system—however well it *simulates*—can know, for example, what it feels like to be humiliated as a human being, because no other system incorporates within itself the experiences of a human being. Yet judgment, a part of human reason, employs, among other things, just such knowledge. The computer metaphor of mind is misleading, he insists, because it points to the mind's instrumental uses only and reflects an instrumental view of being human in the world; and this in turn leads to precisely the kinds of social and communications problems that Habermas has addressed.

It is interesting that another recent addition to our genre, *The Emperor's New Mind* (1989), by Roger Penrose, takes up the same objection from a more mathematical point of view. But while Penrose writes as a scientist skeptical about the finality of our latest theories, Weizenbaum takes a more existential stance. He is concerned not just with the validity of certain claims but what we do with those claims. He worries about the uses of the computer metaphor, but even more about its reflexive effects on those who adopt the metaphor. Unlike most scientists, Weizenbaum frequently emphasizes his role as *teacher*. We are all teachers, he often says; our example counts. And it appears he means not just academics but all human beings as such.

Then why do I include him in my list? In what way does his sort of book contribute to the telling of the grand evolutionary tale? This point is important.

The significant thing is that even Weizenbaum does not question the evolutionary epic in its overall shape. He speaks, in effect, from within that epic. Indeed, his view that human reason is linked to feelings derived from experience harmonizes well with sociobiology, for example. He grants that machines with sufficiently complex internal models of themselves could acquire "a kind of self-consciousness"; he merely wants it understood that *that kind* of consciousness would be nonhuman, the self-consciousness of an organism from another planet, say (Weizenbaum 1976, 210). Penrose too accepts the main plot of the evolutionary epic. Nearly all critics within the scientific community take issue with only some *interpretations* of *pieces* of that plot; hence, in general their objections add weight to the story rather than detract from it, while arguments like Weizenbaum's penetrate so deeply that by force of relevance they more or less *join* the literature that tells that story. In other words, the serious reader who is attracted to the other books on my list but misses Weizenbaum's (or something similar) has not read *broadly* in this literature.

By including partial dissent, I wish also to call attention to the fact that some aspects of the evolutionary epic are still open, still under discussion. Although many members of the core group speak without qualification, often dogmatically, the genre as a whole must be regarded as somewhat fluid at the periphery. There, significant questions of interpretation, though not raised often enough, are still possible. Yet, this possibility (to which I will return) makes the literature of the epic more serious and interesting than would otherwise be the case. It is this, as much as anything, that saves it from being merely an intellectual fad, a passing ideology, the hobbyhorse of philosophically untutored scientists.

How widely such books are read and how great is their influence can be gauged to some extent by looking at the following example. In a major work, *Does God Exist?* (1980), the theologian Hans Kung offers what he takes to be a scientifically cognizant contribution to Christian apologetics. When examining the place of God in natural history, Kung freely accepts the conceptual categories given in the works of Monod, Eigen, Weinberg, and the others. In a section dealing with the possibility of God's direct role in the creation of life, this Catholic theologian feels constrained to make the negative conclusion (642–47), largely on the authority of *Laws of the Game*.[7]

In view of all that has been said here, the answer to the question I posed at the beginning seems embarrassingly simple. Of course: From its earliest moments, science has had for us a profound *cosmological* significance, quite aside from any other interests or uses or appeals; *cosmological,* used here in the Greek sense of "all-encompassing," includes the large, the small, the inner, and the outer. We know this was so in ancient times and during the first scientific revolution; it was evident again in the late nineteenth century; and today once more that significance is increasingly felt, technical difficulties notwithstanding. Yet, during the middle decades of our own century, many commentators tended to overlook this deeply ingrained *cognitive* interest, an interest in the nature of "the whole," as distinct from all technical and aesthetic interests.

Why? The positivist period from which we are now emerging was unique in one respect. By and large, both scientists and philosophers were absorbed by details of fairly limited problems and subdisciplines, emphasizing mechanism and methodology. Quite appropriately, many have called this the "age of analysis." But today, as we witness the new drive for theoretical "unification," we are reminded that an age of analysis is perhaps not typical at all. To the outsider, the larger significance of science could indeed appear marginal during a time when scientists themselves insisted their role was limited to such analyses, and when historians, seconded by philosophers, supported this impression. That is why Aldous Huxley's point about abstraction and the remoteness of science from human life did have an *interim* and partial validity.

Now, however, it is becoming clearer that when analysis is pursued to such great depths, the cognitive significance of science cannot be seen in the fruits of each analytical achievement, expounded singly and in isolation. The spinning ball of methane and ammonia, *in itself,* moves few of us. Whatever images one may derive from it will not impinge on the larger picture of life and of ourselves, regardless of the language used. But if we connect that ball of gas to the evolution of stars, then to the origin of the elements of which our bodies are made, and then to a hidden language of molecular letters that spells our nature, to teleonomy and consciousness— well, that is *another kind* of story.

Contrary to Feynman and Aldous Huxley, the present interest in science's cosmological meaning, and scientists' improved ability to communicate their message, were not spearheaded by the appearance of any great poet or artist. Once again it is the writers' intense *cognitive* involvement with "the nature of things" that spurs on their imagination. It is the cosmological interest that now leads to new, ingenious forms of expression like those of Hofstadter or Eigen and Winkler.

About the scientific view of the world, Aldous Huxley said, "Although it is a determinant of human nature and human behavior, this reality is nonhuman, essentially undramatic, completely lacking in the obvious attributes of the picturesque" (1963, 107). That the new evolutionary epic is "undramatic" or "unpicturesque," none of our authors or their readers would now grant. But from the cognitive point of view, Huxley's omission looms even larger. What he meant by a "determinant" of human behavior, was, for example, atoms, genes, or glandular physiology: the physical stuff itself, technically described. What he clearly did *not* mean was the controlling power of scientific stories *at higher levels of integration*, the power that Wilson, Prigogine, and many others today speak of, the same power that Aldous Huxley's brother, Julian, did notice, well before the present eruption of literature of the third genre.[8]

THE COMMON ERROR AND THE
PLACE OF HERMENEUTICS

Apparently, Habermas's oversight is of the same kind: too restricted a view of the *possible* effects of science on the life-world. Of course, technology has today come into its own as never before; and *mimesis* of a cosmic harmony, though not entirely out of fashion,[9] is, in our man-made environment, extremely difficult. No one understands better than Habermas the role of action-orienting thought structures, dialogically and hermeneutically acquired. Yet the traditional desire to maintain logical separation of the *Naturwissenschaften* from the *Geisteswissenschaften* (or the autonomy of "differentiated value spheres," as Max Weber put it)[10] leads Habermas to assume that natural science cannot seriously "take over the function of meaning-creation" for the "communicative practice" of life (1982, 276). And so, when he reaffirmed that *mimesis* is not an option for modern man, he could only conclude that the *one* channel between cosmos and human self-understanding was closed for good (1971, 302–4).

But now, near the end of the twentieth century, although we do understand that a knowledge of stellar structure cannot affect our self-concept *directly,* that nonequilibrium thermodynamics will not give purpose to life, we are beginning to realize that this is not the end of the matter. For it is not *in themselves* that these bits of knowledge exert their effect; today they do so through the *metaphors* they support, especially the high-level metaphors—extrapolations and integrative appropriations of many disciplines. Through *this* channel the sciences do indeed have action-orienting power. As Mary Hesse puts it, "Society interprets itself to itself, partly by means of its view of nature" (1980, 186). And in a more provocative, insistent way, that is also what E. O. Wilson is saying. That is the basis of his grand view of the evolutionary epic.

In acknowledging Wilson's realistic grasp of what is actually happening between science, literature, and life, we need not accept his *interpretation* of the epic. Here Habermas's instinct to keep this grand vision at arm's length acquires a certain intellectual and moral stature. Habermas is afraid that if it is accepted in the objectivist manner, and if it does in time gain the full power prophesied by its advocates, then the ground of human self-understanding will have been even more drastically misplaced. For then the ways of seeing *from within* the paradigm of this epic, *its* modes of comprehension, will be still harder to confine to the native expert cultures. Arising primarily from instrumental interests, these modes favor an objectivizing approach, which eventually leads to something like "colonization" of the life-world (Habermas 1985, 209; Bernstein 1985, 23). Therefore, to keep communication "unconstrained" by scientistic language, Habermas and many like-minded thinkers wish to preserve domains within the culture where truth is tied to consensus rather than to method.

But if Habermas's basic assumption is wrong—if the recent separation of science from the moral and aesthetic spheres is not final, as he believes—then much depends on *how* these grand myths of science are appropriated, how they fit in with the whole cognitive ecology. And this is just where reflection on meanings, dialogical clarification, and communicative rationality become especially important. By constructing high-level metaphors and extrapolating natural science to the proportions of a "grand myth," the new story-tellers enter the realm of hermeneutics. By impinging on the life-world, not just instrumentally but as socially orienting theory, natural science does indeed blur that boundary between "value spheres" (of which Weber and Habermas speak), and does open itself to modes of discussion usually reserved for the human sciences.

Such a mode, *hermeneutics,* as the term is used here, is the conscious attempt to interpret difficult linguistic or symbolic messages coming from remote or unfamiliar sources, whatever they may be. It requires a circular, bootstrapping type of thought—the "hermeneutic circle"—in which the parts are needed to explain the whole, but also the whole to understand the parts (see Gadamer 1984). Recently, philosophers of science have pretty much come to agree that such "circular" routes to knowledge lie at the very foundation of what we call science. Data are not "raw" and pure; phenomena are not free of theory. Rather, on the basis of phenomena *interpreted* with the aid of older theories, we construct *new* theories that *re*interpret the phenomena. If one grants even as a remote possibility that language is not isomorphic with any natural ontology (Arbib and Hesse 1986, 181) and that physical theories may therefore progress in the sense of pragmatic problem solving without necessarily converging to a "final picture," or without converging in a way we can monitor, then natural science itself, with all its successive models and theories, represents just such a hermeneutic effort: the effort to "read the book of nature."

In the philosophy of science, positions for and against this view are well known.[11] For my purpose, though, the negative arguments (such as those of Markus) are irrelevant because the "objectivations" here, typical of that third literary genre, embody "mythical" extensions and metaphors of high generality. At this level, interpretation is at least implicit; if it is not made explicit, the danger of naïve objectivism looms larger.

When, for example, Eigen and Winkler describe biological or prebiological evolution as a "game," complete with "strategies" (whose?) as well as rules—and they do so not just for the sake of popular exposition but for scientific understanding—they surely are saying something that nature "in herself" is not saying unequivocally to everyone, not even to most biologists. What they do, among other things, is to reinforce a powerful metaphor, the "strategy" of the genes, which itself becomes an object of further scientific theorizing. To treat the world this way is an act of interpretation. A further, compounded interpretation follows, as in the work of the sociobiologists, if one proceeds to build on this metaphor to draw conclusions about human nature. Thus, interpreting evolution in terms of "strategies" leads to an "understanding" of certain differences in human behavior that turn out to be extremely controversial.[12] Such disputes puzzle people unaccustomed to dealing explicitly with the interpretative component of science. But the preceding discussion shows why controversies of this sort should be expected, and why they do not differ much from those in the social realm. A hermeneutic approach here is useful, first, because it calls attention to the interpretative component and to possible alternative interpretations. Second, at a higher level still, it demands a historic understanding of *science as a whole* before particular theories, even the latest and the best, can be truly understood outside the context of immediate technical application.

Again, in regard to the sciences of intelligence and consciousness, the same hermeneutic need arises, only more so. Here, from the very beginning, knowledge and self are inextricably related. The problem is sufficiently difficult that interpretation often comes in layers, metaphors nested in still more general metaphors. A large, complex AI program is pictured as a "bureaucracy" of competing jurisdictions, of courts, or of "little societies" (of subroutines), each with its own character. Not mere professional jargon this; for in time, the same metaphors and submetaphors, willingly adopted, refer to the professional himself—and to us all (Turkle 1984, 287).

It is the merit of the third genre of science literature that, on the whole, it pays attention to the historical perspective, exhibits this tendency of science to generate metaphors at every level, and stimulates examination of these metaphors. When we try to understand, for example, consciousness as self-reference (strange loops), we face squarely the twin insights that science is *metaphor* and science is *self-knowledge*. How are such metaphors to be understood *if* their reflexive, self-characterizing function is to be appropriated? Note how crucial to this issue is the *purpose* of the appropriation. In the context of scientific work on computers, the question "How are such metaphors to be understood?" either has no place at all or it has an entirely different meaning with a relatively straightforward answer. Only in an educative context of enlightened cultural concern, the context assumed by the authors of that third genre of science literature, does interpretation of metaphors become a central task. In an informal way, this precisely was Weizenbaum's task in regard to the computer metaphor of mind. By critically examining the personal and social implications of this metaphor, as well as the science behind it, its development, and its effects, he clarified substantially its *meaning for us*.

It is not true that hermeneutics applies only to the "human sciences." It applies to any interpretative science. No doubt, natural science in its most formal modes is least interpretative; but wherever it *is* interpretative, there hermeneutics is relevant too. There we are once again dealing with humanly constructed texts, their metaphors and high-level theories, which come to us with a context and a history.

NOTES

1. Yet Stephen Gould, for example, sees no essential difference between his strictly scientific writing and his "popular" essays—only the technical arguments being omitted from the latter (lecture at Worcester Polytechnic Institute, Worcester, Mass., 8 October 1987).

2. Particularly worth noting are Yantsch (1981), Davies (1983), Sperry (1985), Chaisson (1981, 1987), Margulis and Sagan (1986), and Penrose (1989). These works fit my description in varying degrees. Chaisson's books are far less demanding than *Gödel, Escher, Bach*, for example, while Yantsch's contribution is a collection of papers presented at a meeting of the American Association for the Advancement of Science.

3. E. O. Wilson, for example, mentions "the Huxleys, Waddington, Monod, Pauli, Dobzhansky, Cattell" (1978, 214).

4. On Bohr's and Pauli's views, see Mayr (1982, 428–29). On Wittgenstein, see Putnam (1981, 108–9). Some of Popper's questions are found in his *Objective Knowledge* (1972, 267–70).

5. Haeckel, for example, gives many qualitative discussions of developments that have only recently turned quantitative; for example, the size of proteins ("plasm molecules"), prebiotic evolution, and what is now known as the Gaia hypothesis— the idea that the whole earth should be regarded as one giant organism (1905, 36).

6. Many scientists object to Wilson's, and others', use of the word *myth* in this way. From the viewpoint of the philosophy of science, I prefer *metaphor* or *high-level metaphor*.

7. Further evidence of the effects of this literature can be seen in *Zygon* 24(2) (1989), where the authors in my sample are much quoted in a debate on science and the theology of Wolfhart Pannenberg.

8. Julian Huxley's view of the larger sociomoral role of science is very close to that of Monod and Wilson. See, for example, J. Huxley (1957), especially the essays "Man's Place and Role in Nature" and "Evolutionary Humanism."

9. For a modern version of living in "harmony with nature," see Toulmin (1982, 262).

10. The idea of the "differentiation of value spheres" originates with Max Weber and is now used extensively by Habermas, as in his *Theory of Communicative Action* (1984, 233–47).

11. Proponents of hermeneutics in natural science are in the minority. But Hesse (1980), Arbib and Hesse (1986), Grene (1985), and Stent (1985) represent this position well. Anthony Giddens and Habermas have countered that nevertheless, social science is different because it has an *additional* need for hermeneutics: scholars must first use interpretative understanding to gain access to a language already in use, prior to any theory construction (Habermas 1984, 109–11); only after that do their problems begin to resemble those faced by natural scientists. Such con-

siderations do not undermine my argument here that in natural science at least one important role for hermeneutics exists. However, one could go further. The outsider, the nonscientist, the student, and sometimes even the researcher may have need for the "double hermeneutic" as well: first, to "gain access" to the logic of a science *as a whole* (through history, philosophy), and only then to *properly* understand particular theories in light of the evidence. Many scientists would not accept this, and to argue it here would be a digression. But the recent attention given by biologists to the history and philosophy of evolution, as this theory undergoes yet another reappraisal, is some indication that my point has support.

12. In 1989, Philippe Rushton, a professor at the University of Western Ontario, was the target of protests and of investigations by Ontario police for publishing scientific papers that some people describe as racist and reactionary. These papers suggest that certain behavioral differences between major ethnic groups can be traced to different evolutionary strategies adopted by these groups. See Gross (1990).

REFERENCES

Arbib, M. A., and M. B. Hesse. 1986. *The Construction of Reality*. Cambridge: Cambridge Univ. Press.
Bernstein, R. J., ed. 1985. *Habermas and Modernity*. Cambridge: MIT Press.
Chaisson, E. 1981. *Cosmic Dawn*. Boston: Atlantic–Little, Brown.
———. 1987. *The Life Era*. New York: W. W. Norton.
Chambers, R. 1845. *Vestiges of the Natural History of Creation*. 2d ed., from the 3d London ed. New York: Wiley and Putnam.
Davies, P. 1983. *God and the New Physics*. New York: Simon and Schuster.
Eigen, M., and R. Winkler. 1981. *Laws of the Game*. Trans. R. Kimber and R. Kimber. New York: Harper and Row.
Feynman, R. P., R. B. Leighton, and M. Sands. 1963. *The Feynman Lectures on Physics*. Vol. 1. Reading, Mass.: Addison-Wesley.
Fiske, J. [1874] 1902. *Outlines of the Cosmic Philosophy*. 4 vols. In *The Miscellaneous Writings of John Fiske*. Boston: Houghton Mifflin.
Gadamer, H. G. 1984. *Truth and Method*. Trans. G. Barden and J. Cumming. New York: Crossroad.
Grene, M. 1985. Perception, Interpretation and the Sciences. In *Evolution at a Crossroads*. Ed. D. J. Depew and H. Weber. Cambridge: MIT Press.
Gross, B. 1990. The Case of Philippe Rushton. *Academic Questions* 3: 35–46.
Habermas, J. 1970. *Toward a Rational Society*. Trans. J. Shapiro. Boston: Beacon Press.
———. 1971. *Knowledge and Human Interests*. Trans. J. Shapiro. Boston: Beacon Press.

————. 1982. A Reply to My Critics. In *Habermas: Critical Debates*. Ed. D. Held and B. Thompson. Boston: MIT Press.

————. 1984. *The Theory of Communicative Action*. Vol. 1. Trans. T. McCarthy. Boston: Beacon Press.

————. 1985. Questions and Counterquestions. In *Habermas and Modernity*. Ed. R. J. Bernstein. Cambridge: MIT Press.

Haeckel, E. 1905. *The Wonders of Life*. Trans. J. McCabe. New York and London: Harper.

Hesse, M. B. 1980. *Revolutions and Reconstructions in the Philosophy of Science*. Bloomington: Indiana Univ. Press.

Hofstadter, D. R. 1979. *Gödel, Escher, Bach: An Eternal Golden Braid*. New York: Vintage Books.

Huxley, A. 1963. *Literature and Science*. New York: Harper and Row.

Huxley, J. 1957. *Knowledge, Morality and Destiny*. New York: Mentor Books.

Kierkegaard, S. 1954. Sickness unto Death. In *Fear and Trembling and Sickness unto Death*. Garden City, N.Y.: Doubleday.

Kung, H. 1980. *Does God Exist?* Trans. E. Quinn. New York: Doubleday.

Margulis, L., and D. Sagan. 1986. *Microcosmos*. New York: Simon and Schuster.

Markus, G. 1987. Why Is There No Hermeneutics of Natural Science? *Science in Context* 1: 5–51.

Mayr, E. 1982. *The Growth of Biological Thought*. Cambridge: Harvard Univ. Press.

Monod, J. 1971. *Chance and Necessity*. Trans. A. Wainhouse. New York: Random House.

Penrose, R. 1989. *The Emperor's New Mind*. New York: Oxford Univ. Press.

Popper, K. 1968. *Conjectures and Refutations: The Growth of Scientific Knowledge*. New York: Harper and Row.

————. 1972. *Objective Knowledge*. Oxford: Clarendon Press.

Prigogine, I., and I. Stengers. 1984. *Order out of Chaos: Man's New Dialogue with Nature*. New York: Bantam Books.

Putnam, H. 1981. *Reason, Truth and History*. Cambridge: Cambridge Univ. Press.

Rose, S. 1976. *The Conscious Brain*. New York: Vintage Books.

Simpson, G. G. 1949. *The Meaning of Evolution*. New Haven: Yale Univ. Press.

Spencer, H. [1880] 1958. *First Principles*. 4th ed. New York: De Witt Revolving Fund.

Sperry, R. 1985. *Science and Moral Priority*. New York: Praeger.

Stent, G. 1985. Hermeneutics and the Analysis of Complex Biological Systems. In *Evolution at a Crossroads*. Ed. D. J. Depew and H. Weber. Cambridge: MIT Press.

Toulmin, S. 1982. *The Return to Cosmology*. Los Angeles: Univ. of California Press.

Turkle, S. 1984. *The Second Self*. New York: Simon and Schuster.

Weinberg, S. 1979. *The First Three Minutes*. New York: Bantam Books.

Weizenbaum, J. 1976. *Computer Power and Human Reason*. San Francisco: W. H. Freeman.

Wilson, E. O. 1978. *On Human Nature*. New York: Bantam Books.

Yantsch, E., ed. 1981. *The Evolutionary Vision*. Boulder: Westview.

Ideology and

Culture

Nature and Nation in Popular Scientific

Narratives of Polar Exploration

BARRY PEGG

At two points in the "heroic" period of polar exploration, one at its peak in the middle of the nineteenth century and the other marking its close on the eve of the First World War, two scenes present stark contrasts between survival and tragedy. The first scene, probably in the spring of 1846 (Beattie and Geiger 1987, 38), has been only sketchily reconstructed from eyewitness accounts. British sailors from an expedition attempting the Northwest Passage to the Pacific, wearing poorly insulated clothing (tight canvas jackets and leather boots), drag twenty-foot boats over the ice of Victoria Strait, in the vicinity of the North Magnetic Pole, toward what they think of as civilization. They are dying—starving, freezing, and, as we now know, disoriented from lead poisoning due to carelessly soldered canned provisions from London (Beattie and Geiger 1987, 156–60) and in an advanced stage of scurvy (Beattie and Geiger 1987, 56). Lieutenant Francis Crozier, leading these forty survivors after the death of the expedition's leader, Sir John Franklin, begs food—both by sign language and using what he thinks is the Eskimo word for *seal*—from four Eskimo families. The Eskimos, by contrast, are better fed because they know how to hunt the sparse game of the region and have food traditions that preserve maximum nutrition, even scarce vitamin C. The next morning, the Eskimos seemingly reverse their hospitable treatment and depart, leaving the sailors to their fate. It is probable that "the thin resources of that part of the Arctic would not support the four families and a party of forty men, and the Eskimos knew it" from tradition and experience (Lopez 1986, 380).

Sixty-six years later, at the other end of the world, another scene of starvation and failure is played. Attempting to return from the South Pole, Captain Robert F. Scott and his two remaining companions, Lieutenant Henry Bowers and Dr. Edward Wilson, also starving and freezing and probably, like Franklin's men, suffering from scurvy, are awaiting their deaths in a blizzard-engulfed tent eleven miles south of a depot that would have saved them. Their sledge, which they have been "man-hauling" on foot, carries thirty extra pounds of geological specimens, among them a plant fossil which is later to play an important part in "building the . . . originally derided . . . theory of continental drift" (Young 1980, 16). In this scene, the players who know how to deal with the harsh environment are not present except in the minds of the dying men: Roald Amundsen and his "Norskies," as the British disparagingly called their rivals, have raised the Norwegian flag at the South Pole about a month before. The Norwegians' dog and ski tracks might have been covered by drifting snow by the time the British party arrived at the pole (there is no known record of the British mentioning them), but those tracks were the marks of a travel technology, rejected by the British party, that carried the Norwegians to the South Pole and back safely, expeditiously, and in good health.

What do these two scenes have to tell us about the interplay of cultures in the planet's two coldest environments? What do they tell us about what *nation* means in an environment whose inhabitants, like many unvisited cultures of the recent past, call themselves just "the people"? And, more important for the present purpose, what do the popular scientific narratives that chronicle those two expeditions, and others like them, have to tell us about how the British, Norwegian, and American cultures withstood and interacted with an indifferent nature and an uncomprehending indigenous culture?

This essay attempts to answer those questions by examining the interaction of national pride with the exploration of the natural world and its inhabitants, as it is seen in the popular science writing of the period; that is, in the accounts of the expeditions as edited for the public. I survey both the generation of information and its inhibition. On the one hand, knowledge was generated from contact with skilled practitioners, improvisation, empiricism, and reading; but, on the other hand, various types of cultural resistance—linguistic, institutional, commercial, geographical, ecological, and imperialistic—held back its application. The popular scientific accounts show these contrary processes sometimes predominating in different expeditions, sometimes mixed together in the same one.

What exactly was the role of popular scientific narratives in describing, rejecting or accepting, and passing on polar travel and survival technologies to the succession of explorers? We could start by looking at what these scientific narratives include: both information about what anthropologists call the "material culture"—the techniques of daily life, the life-support system—so starkly absent in the cases of the Franklin expedition on King William Island in 1846 and the Scott party on the Ross Ice Shelf in 1912, and clarification of the role of nature and nation in spreading and inhibiting these technologies.

Clothing, shelter, and diet are all crucial where life is as fragile as a candle flame in a storm. The absence of vegetation means that vitamin C, if it is to be obtained from the environment, must be secured from raw meat, and the cold dictates a diet very heavy in fat. The need to hunt animals that have a blubber layer means alternating periods of waiting (typically, kneeling on the ice by a breathing hole) with bursts of very heavy work like hauling or butchering. Hunting this way necessitates clothing that is very warm but loose enough to allow sweat to evaporate during periods of activity, because wet skin loses heat very quickly. Amundsen's *North West Passage* is one account that passes on the discovery of this technique (1908, 1:45).

The need to travel long distances in search of game requires (literally) cold calculation; somehow or other through prehistory a technology was refined in which the dog provided motive power, aid in hunting, direction finding in blizzards, and, in an emergency or on exceptionally long trips, food for other dogs or for drivers. The need to travel light (because excess weight might mean the difference between reaching home or dying on the return trip if food ran out) meant that igloos or snow houses built on the spot and efficiently heated by animal-fat lamps were preferable to heavy skin tents. John Rae learned this in the 1830s by emulating the Eskimos (Berton 1988, 158–59); Amundsen recounts how he learned the technique (1908, 1:125–27); and Robert Peary built up a relationship with an Eskimo village to supply him with dogsled drivers and igloo builders (1986, 43). Travel on foot could be improved on, though Arctic peoples did not know this: skis, developed prehistorically in Scandinavia, were brought to the Arctic for Fridtjof Nansen's first Greenland crossing in 1888, and combined with dogsleds and taken to the South Pole by Amundsen. The popular scientific narratives often illustrate how nature (the climate, terrain, and fauna) dictates the culture of a region. To the Eskimos of the nineteenth century (following Lopez [1986, 418], I use this term rather than the geographically restricted *Inuit*), hunting was either the purpose of travel or the usual means of fueling people and dogs on a journey. The climate and terrain precluded major investment of energy in purely scientific activities.

The Europeans, on the other hand, were required by their scientific culture to demonstrate or seek mastery over nature—both national pride and commercial fortunes depended on it. Even now it is hard for us to see the poles as anything but arbitrary points on the map, suitable for a demonstration of prowess like the finish line of a race or a goal in a team sport. Such motives must have been quite without meaning to nomadic hunters like the Eskimos; such objectives must have looked like simple recklessness. The only point of the trip for the Eskimos who accompanied Peary and Matthew Henson on the series of six northern trips culminating in the 1909 North Pole attempt, for instance, must have been the firearms and sled materials Peary paid them with (Peary 1986, 333).

In the popular scientific narratives, these survival implements, techniques, and traditions—Eskimo in the Arctic and a hybrid of Eskimo and Scandinavian in the Antarctic—take center stage again and again as the subjects of a dialogue between cultures. In many cases the dialogue takes the following pattern: the indigenous material and spiritual culture appear as the subjects of ethnographic writing by the travelers, and the writers and editors of a popular scientific narrative reveal (sometimes unconsciously) their cultural assumptions and prejudices in response to them.

Popular science books were the medium for reporting this cultural interface—at least the "civilized" side of it. Apart from newspaper journalism (incidentally a field ripe for research), the public was informed about the expeditions by means of two- or three-volume books. Orders, logs, and notebooks written in the field were edited for popular palatability and appeared with increasing frequency following the three Parry expeditions (between 1819 and 1825). Only the output of expedition leaders, as opposed to their companions, appeared before World War II, however, and then only in heavily edited form. In the case of *Scott's Last Expedition* and Amundsen's *South Pole*, papers by scientific members appeared as appendixes, usually in a second volume, representing an attempt to justify the scientific value of the expeditions, and to avoid accusations of mere nationalism. These papers report auxiliary expeditions, geology, flora and fauna, and the like. Rather than the short popular accounts in magazines we might expect today, these accounts, relatively undemanding in literary style and technical content, served the same public function as *Discovery*, *Scientific American*, or *Science News* do today, and corresponded to some extent to scientific reports in today's learned journals.

The style of the main narratives is personal, anecdotal, and motivational. Scott's narratives in particular, and Amundsen's to almost the same extent (*pace* Huntford's attempt to contrast Scott's "romantic excogitation" with Amundsen's "sailorlike simplicity" [1979, 191, 192]), were crafted to give an impression of stoutheartedness under adversity (with the addition of ingenuity in the case of Amundsen's account). The narratives and their slightly more scientific appendixes fall within Greg Myers's category of the "narrative of nature" rather than the "narrative of science" (1990, 144–64)—that is, they recount phenomena chronologically and from the investigator's point of view rather than according to the topical structure of the subject.

I believe that a particular expression of a national sense of superiority, namely the British navy's prejudice, must be blamed for the initial resistance to learning the technology needed for polar travel and survival. The British navy went into exploration for the wrong reasons in the first place. After Napoleon, "the Navy had to find something for its ships, its men, and, most important, its officers to do now that Europe was at peace" (Berton 1988, 18). The navy brought an entrenched arrogance to its new task, even discounting the opinions of experienced navigators, as if only navy men could know the truth. The most experienced sailor of his time was the whaling captain William Scoresby, whose *Polar Ice* (1817) has since been called "the foundation stone of Arctic science" (Berton 1988, 25). But—he wasn't navy.

Berton also sums up neatly the attitude of the Royal Navy to the Eskimos at the time of the first Parry expedition of 1818: "nobody on this so-called scientific expedition thought to investigate how a band of people who couldn't count past ten had managed to adapt to their formidable homeland" (1988, 30). This in spite of the fact that the Eskimos were expert mapmakers and showed Lieutenant William Parry the existence of a passage now called Fury and Hecla Strait (Berton 1988, 51). The Ross and Parry party of 1818 greeting the Eskimos of Etah, a remote settlement in northern Greenland, in gold braid and cocked hats, as if they were attempting to make a good impression at the court of Versailles, could well serve as a snapshot characterizing the period of exploration leading up to the disastrous Franklin expedition of 1845–47.

Hence, infrequent contact and cultural barriers prevented the British from taking advantage of Eskimo techniques. Berton sums up the prevailing attitude of nineteenth-century Britons to "native" cultures very well:

The nineteenth-century English upper classes . . . considered themselves superior to most other peoples, whether they were Americans,

Hottentots, or Eskimos. But another part of it, surely, was fear: the fear of going native. Could any proper Englishman traipse about in ragged seal fur, eating raw blubber and living in hovels built of snow? Those who had done such things in some of the world's distant corners had been despised as misfits who had thrown away the standards of civilization to become wild animals. Besides, it was considered rather like cheating to do things the easy way. The real triumph consisted of pressing forward against all odds without ever stooping to adopt the native style. (1988, 8)

Occasionally, as with George Francis Lyon, an officer in Parry's 1821–22 Arctic expedition, there were exceptions. Lyon lived with the Eskimos and learned their ways and foods (including *nerooka*, a delicacy consisting of the contents of the entrails and stomachs of deer—not for the squeamish but certainly high in vitamin C). He did this, as his contemporaries read in his published journal, "on the principle that no man who wishes to conciliate or enquire into the manners of savages should refuse to fare as they do" (quoted in Berton 1988, 50).

The resistance to Eskimo culture evident in the major popular scientific narratives reflects the way Europe since Napoleon and the United States since the Civil War had come to regard themselves as the civilized part of the world, destined to push back the unenlightened darkness of the rest of the globe. France, Britain, Portugal, Belgium (the most notoriously corrupt African colonial power, as seen in Joseph Conrad's *Heart of Darkness*), and the new Germany were claiming Africa; even Denmark had laid claim to Greenland. Against this imperialistic background Norway achieved independence from Sweden on June 7, 1905, and a new nonimperialistic kind of national pride was embodied by Fridtjof Nansen, Norway's first ambassador to London. In 1895 he had come 170 miles nearer to the North Pole than anyone else, and later he served as a model of Saint Olav for a famous artist's illustrated edition of the Sagas, entering "thousands of Norwegian homes in the likeness of a medieval Norse hero and the patron saint of Norway" (Huntford 1979, 54). His sense of the heroic can be seen in his introduction to Amundsen's *North West Passage*, where he writes of "a ring of steeled, purposeful human will—through icy frosts, snowstorms, and death" (1908, xxiv–xxxv).

Nansen's *First Crossing of Greenland* (1890) is remarkable not only for its nationalism but also for its empiricism. Nansen recounts how he improved on the rather inconclusive British naval sledging experiments with thin metal-shod runners (Cyriax 1963, 129–30). "We Norwegians," Nansen says in an interesting conflation of nation and nature, "look upon this expedient [broad runners on sledges] as simply natural, as we are accustomed to our old-fashioned 'skijaelke,' which is a low hand-sledge on broad runners, resembling our ordinary ski" (1890, 1:33). Nansen followed the Swedish explorer Nils Nordenskjöld in the use of skis in the Arctic, having been impressed by the "extraordinarily long distance [covered by Baron Nordenskjöld] in an astonishingly short time" on skis. Nansen also reasoned out a simultaneous solution to both the motive power and food supply problems in an empirical, even calculating, way. He first proposed using either reindeer or dogs for motive power and food, and then narrowed the selection down to dogs on the grounds that they are easier to feed (they eat the same food as humans, while reindeer have to have reindeer moss brought along); and though their flesh takes some getting used to, "the Eskimo reckon it a delicacy, and . . . anyone who could not bring himself to eat it would not be a fit person to accompany such an expedition at all" (1890, 1:30). In sum, the nationalism in Nansen's scientific narrative is not so strong that new ideas cannot be entertained.

But Nansen is only one of the notables of the so-called heroic period. The accounts of Scott (1905, 1913), Peary (1910), and Amundsen (1908, 1912) reveal three different kinds of interplay of nature and nation. Scott's shows the failure of blinkered nationalism in the Antarctic; Peary's shows how his "American plan" followed Amundsen's and Cook's Arctic examples of learning from the Eskimos, but betrays an inability to accept nonwhites as anything but primitive; and Amundsen's combines nationalism with receptivity to information from others.

Scott's Antarctic expedition of 1901–4 is recounted in *The Voyage of the "Discovery"* (1905). This popular two-volume scientific memoir, whose great success was a factor in the funding of his 1911–12 expedition, was designed for the general reader. Poetic epigraphs for each chapter (in spite of a claim to tell the tale "as simply as possible" [1905, 1:viii]) appeal to the parlor. A rather ignoble appeal to jingoism appears in the historical preamble, in which Scott makes Eurocentric fun of the South American Quiros who "solemnly" annexed a mistaken Antarctic (1905, 1:5). In general the account betrays a reluctance to experiment with unfamiliar technology. Even though "Norwegian snowshoes or ski" (1905, 1:125) were brought along, for example, they were used mainly for exercise, as the caption to a group photograph suggests (1905, 1:facing 300). Scott thus failed to benefit from the available accounts of Nansen, Amundsen, and Peary, who had progressively built up a tradition of skis and dogsleds as the most energy-efficient polar travel method. While Scott's eventual preference for man-hauling was based on empiricism, unfortunately his perfunctory "test" was conducted in ignorance of skiing technique. He found that "a party on foot invariably beat a party on ski even if the former were sinking ankle-deep at each step; while to add to this, when the surface was hard, ski could not be used." Following the need to do things the hard way characteristic of all British naval expeditions, Scott thought skis good only for "a party out of condition." It is quite true that it is difficult to get skis to glide at very low temperatures (a condition modern Nordic skiers still find it tricky to wax for), but still it must be said that Scott allowed the inadequacy of his ski expertise to dictate a travel method inordinately hard on his party, though he tried to make a virtue of necessity with grand phrasemaking: there is, he says, "nothing to equal the honest and customary use of one's own legs" (1905, 1:454).

As Scott's *Discovery* expedition was ending, Amundsen, so he tells us in *The North West Passage* (1908), was attempting to learn from the indigenous people of King William Island the kind of techniques Scott was making light of. This meant, for example, learning the difficult language of the Netsilikmiut (Seal Eskimos) among whom he wintered on the Northwest Passage expedition of 1903–7 (1908, 1:291). His willingness to learn an unfamiliar tongue suggests that he viewed his own culture as less than monolithic, as is further shown in his account of the preliminaries to the first contact with his native mentors. His narrative (as edited for publication) demonstrates a surprising degree of cultural self-parody:

> Down on the ice I drew up my troops and inspected them, and even the most critical general could not have found fault with their appearance and bearing. I myself threw out my chest as well as I could, drew myself up, made a regulation right-about turn, and gave the command "Forward—march!" . . . I advanced, casting a sidelong glance up to the deck where the Lieutenant and the cook stood side by side. It seemed to me that their expression . . . was not exactly one of admiration, not even of seriousness. (1908, 1:116)

Amundsen does not lose the opportunity to reveal one further irony of all this military deployment. The Eskimos' response to the Norwegians' nervous, rehearsed "Teima" is an even warmer one—"Manik-tu-mi!—the Eskimo's friendliest greeting" (1908, 1:116).

Amundsen also describes how he learned to avoid man-hauling, after he had had to do it on one trip "because snow too cold and too few dogs. By offering all our strength, we managed to advance 3 ½ miles," he notes in his log (quoted in Huntford 1987, 27b). From that moment Amundsen feared man-hauling and learned to depend on dogs. In spite of his response to native culture, however, when concepts of property later became a problem, Amundsen felt obliged to blow up an empty igloo at a distance in a terrorist display of explosive European technology to make the Eskimos think that he had magic powers to punish them (1908, 1:260).

Peary, too, recognized that travel in high latitudes depends on using the techniques of those who live there. In *The North Pole* Peary writes, "It seems unnecessary to enlarge on the fact that the men whose heritage is life and work in that very region must present the best obtainable material for the personnel of a serious Arctic party" (1986, 6). In pursuit of his "American plan," Peary painstakingly built up strong ties with Etah, on Smith Sound in northwest Greenland, the Eskimo village closest to the North Pole. Even so, his tone often reveals condescension and even arrogance: "From the very beginning of my polar work I believed that these most northerly human beings in the world could afford me invaluable assistance in my plans for exploration. Later I had a fatalistic feeling that the Almighty had put the little tribe in this particular place for the express purpose of assisting [me] to win the pole" (1986, 179).

Peary's "American plan" brought him to the same conclusion about "motors" as Amundsen: ponies weigh ten times as much as dogs, he observes in his *Secrets of Polar Travel*, so "every motor that one [pony-drawn] expedition loses means a loss of ten per cent of its tractive force, while every motor that the other [dog-drawn] loses means only one per cent loss" (1986, 198). Dogs were also a better bet than machinery:

> Devices which work satisfactorily in temperate regions are more than likely to fall down when called upon to perform under the handicap of polar conditions. Sooner or later—and usually sooner—any machine will fall down in polar work, and when it does so it is simply a mass of old junk which neither men nor dogs can eat, and which cannot even be burned to cook a pot of tea. (1917, 197)

Peary seems to have been a goal-oriented and persistent man. Anyone who gets within thirty miles of the North Pole at the age of fifty-six, toeless from frostbite, deserves that much at least. Hence it is reasonable to assume that his acceptance of Eskimo culture was factitious rather than sentimental, and that he carried the expected arrogance of the civilized toward the supposed primitive. His attitude to his African-American assistant, Matthew Henson (who just may have been the first human to reach the North Pole), is poignantly instructive: "While faithful to me, and when *with me* more effective in covering distance with a sledge than any of the others [Henson was a better sledge driver even than the Eskimos], he had not, as a racial inheritance, the daring and initiative of [the other, white members of the expedition] Bartlett, or Marvin, or MacMillan, or Borup" (1986, 273). And even though Marvin turned back just before the penultimate stage, Peary writes that "with the exception of Bartlett and myself, [Marvin] alone of *all white men* had entered that exclusive region which stretches beyond 86 [degrees,] 34 [minutes] north latitude" (1986, 254; emphasis added).

Henson was listed in the expedition's roster as an "assistant," eighth in the list of twenty-two crew, after the three young white "assistants" but before the mate and the rest of the ship's crew. An exceptionally capable and ingenious man, he was certainly invaluable, for example, in "training the new men in the art of dog driving, igloo building, and survival" on arrival at Cape Sheridan (Ferris 1989, 56). Still, Peary seems not to have considered Henson a full member of the party. The illustrations to *The North Pole* (1986, between 16–17) show all the crew except Henson in a series of nine portraits. Later pictures include a striking madonna-and-child picture (the baby is probably Henson's son, Ahnahkaq; see 80 Years Later 1987, 25), but Henson himself appears only once, in the center of the Eskimo group posing as standard-bearers at their goal. The fact of the matter was probably that Peary regarded Henson and the Eskimos as on a par.

The same paternalism can be seen when Peary writes about intervening in the lives and livelihoods of the Etah Eskimos to suit his own ends. Cultural mingling, for him, seems to have been trading his goods and his planning for native expertise and endurance.

> Since 1891 I had been living and working with these people, gaining their absolute confidence, making them my debtors for things given them, earning their gratitude by saving, time after time, the lives of their wives and children by supplying them with food when they were on the verge of starvation. For eighteen years I had been training them in my methods; or, to put it another way, teaching them how to modify and concentrate their wonderful ice technic and endurance, so as to make them useful for my purposes. (1986, 43–44)

Peary's reasons for leaving Eskimo culture untouched seem to have been based on a sort of apartheid rather than on true appreciation. He says that the absence of the concept of private property is effective in their environment, but he certainly implies it is inappropriate for European cultures.

> I hope no efforts will ever be made to civilize them. Such efforts, if successful, would destroy their primitive communism, which is necessary to preserve their existence. Once give them an idea of real-estate interest and personal-property rights in houses and food, and they might become as selfish as civilized beings; whereas now any game larger than a seal is the common property of the tribe and no man starves while his neighbors are gorging themselves. If a man has two sets of hunting implements, he gives one of them to the man who has none. It is this feeling of good-fellowship which alone preserves the race. I have taught them some of the fundamental principles of sanitation and the care of themselves, the treatment of simple diseases, of wounds, and other accidents; but there I think their civilization should stop. (1986, 71)

Even so, Peary did not hesitate to intervene drastically, albeit with good intentions, in their economy: "We left them all better supplied with the simple necessities of Arctic life than they had ever been before, while those who had participated in the sledge journey and the winter and spring work on the northern shore of Grant Land were really so enriched by our gifts that they assumed the importance and standing of Arctic millionaires" (1986, 333).

The Eskimos' qualities, for instance their endurance and their formidable survival skills, Peary usually explains (or rather, dismisses) in terms of instinct or nature, as if they were animals or children. "My Eskimos know who built or even who has occupied an igloo, with the same instinct by which migratory birds recognize their old nests of the preceding year; and I have traveled these arctic wastes so long and lived so long with these instinctive children of Nature that my sense of location is almost as keen as their own" (1986, 311–12). Peary, it seems, was capable of looking on their skills as instinct but on his own as experience, and their technology as childlike but his own as civilized.

The North Pole clinches Peary's cultural arrogance in two areas where his (and our) culture and that of his Eskimo guides and porters showed important differences: death and ingenuity. In the first case, cultural patterns regarding death were at work that Peary simply didn't understand, and in the second, while professing admiration for Eskimo culture, he couldn't avoid condescending to an "inferior race" (1986, 333). His reaction in *The North Pole* to Eskimo death practices reflects the European repugnance to Eskimo ways. He was probably more or less used to their odor and table manners and understood (or had modified) their traditional concept of communal vicinity property, but he balked somewhat at their ideas of death. Their apparent callousness was incompatible with the sentimental Christianity of the time. As Peary puts it, "even a mother who has been inconsolable at the death of her baby soon laughs again and thinks of other things" (1986, 67).

When Ross Marvin, a young and promising civil engineer, was drowned at the Big Lead (a strip of open water of changing size and position about forty miles from land), the two Eskimos who were with him stayed behind to break camp and found his floating body later. Peary writes that

> of course they knew what had happened to Marvin; but with childish superstition peculiar to their race they camped there for a while on the possibility that he might come back. But . . . when he did not come back, . . . they were in dread of his spirit. So they threw from the sledge every thing they could find belonging to him, that the spirit, if it came back that way, might find these personal belongings and not pursue the men. (1986, 319)

It seems Peary did not know, or was ignoring, the custom of burying belongings with the dead, disordered in this case by the danger of thin ice and the sense of being on a pointless journey in terrain devoid of game.

Peary did show admiration for Eskimo ingenuity, but in a rather roundabout way: "My own experience has been that the average aborigine is just as content with his own way as we are with ours, just as convinced of his own superior knowledge, and that he adjusts himself with his knowledge in regard to things in the same way that we do" (1917, 44). In *The North Pole*, when promoting the use of his own equipment for his own purposes, he is even more remarkably patronizing about the Eskimos' ingenuity.

> They exhibit, . . . in marked degree, all the Oriental capacity for imitation. Out of walrus ivory, in some respects their substitute for steel— and a *surprisingly* good substitute it is—they will construct *amazingly* good models or copies of various objects, while it does not take them long to master the use of such tools of civilization as may be put into their hands. It will easily be seen how valuable and useful a quality this has proved for the purposes of the arctic explorer. If he could

not rely on the Eskimo to do *the white man's work with the white man's tools*, the labors of the arctic traveler would be tremendously increased. (1986, 61–62; emphasis added)

Both Roald Amundsen and Captain Robert Scott were preparing major expeditions when Peary's account appeared in 1910, and their scientific narratives show the different ways they dealt with Peary's information. *Scott's Last Expedition* (1913), as edited by Leonard Huxley, shows a blindness to all precedents and influences not compatible with naval practice; while Amundsen's *South Pole* (1976) reveals an acceptance of indigenous travel methods, which are then pressed into the service of a nationalism whose chief characteristics are flexibility and independence rather than domination. Amundsen, using Eskimo-derived dogsleds and Scandinavian skis, reached the pole with a small group and returned safely in 99 days. Scott's polar party, on the other hand, traveling in the same season and from the same quarter of the compass but with an assortment of motor sledges that wouldn't start reliably, horses that could neither browse nor walk, skis and dogs that the party only minimally trained themselves to use, and relying mostly on "man-hauling," all in slightly better weather, ended their 103-day trip one hundred miles from base, freezing and starving to death while waiting out a blizzard with insufficient fuel and food.

Amundsen, like Nansen, was certain of the advantage of his Norwegian birth, writing of "us, who were born and bred with ski on our feet," but he did not allow his nationalist tradition to dictate his choice of equipment. Not satisfied with the equipment bought on dry land, he tells us in *The South Pole* that "we spent the winter in altering our whole equipment, which our depot journeys had shown to be too heavy and clumsy for the smooth barrier surface" (1976, 1:viii–ix). In common with most other expeditions of the period, Amundsen relied on tents, although he was led by serendipity and improvisatory genius to rediscover what Eskimos had always known: when you want to build *in* the snow, build *with* the snow. After a big snowdrift had piled up because they had insufficient shovels, "one of us [probably Jørgen Stubberud, the carpenter] had the bright idea of taking Nature in hand, and working with her instead of against her" and dug out "a whole under ground village" (1976, 1:269–70).

Scott's Last Expedition repeats the mix of inspiration and popularization of Scott's earlier book, *The Voyage of the "Discovery."* Edited to conform to the imperial myth, what Roland Huntford calls "an affair of heroism for heroism's sake" (1979, 559), its two volumes were conceived as a memorial in the spirit of Tennyson's "In Memoriam," the Albert Memorial, or indeed the quotation from Tennyson's "Ulysses" used on the Scott cairn near what is now McMurdo Base: "To strive, to seek, to find, and not to yield." Publication had to await the discovery of the bodies and records in the polar spring (i.e., November 1912); the book includes Scott's famous last "Message to Public," in which he attributes to "bad weather" what was really inadequate laying down of supplies. "This journey," he writes, "has shown that Englishmen can endure hardships, help one another and meet death with as great a fortitude as ever in the past" (1:417).

Scott's Last Expedition attempts to conceal, and in doing so reveals, what was worst in the British navy's approach to exploration. Since the navy's purpose was to identify daring and successful middle-level officers, not to reach any particular goal, perseverance in the face of disaster was admired regardless of the outcome. To make matters worse, Scott himself was not a good leader. He left things until the last minute, took criticism very badly, and did not follow the ample documentary advice available to him. His greatest skill was convincing and graceful writing in a grand rhetorical style, unfortunately no substitute for prudence, and indeed detrimental to it.

We can detect in Scott's narrative the bemused puzzlement of someone encountering unfamiliar techniques that challenge the conventional wisdom of the naval college. He writes wistfully of the survival technology he knew of but did not use: Amundsen, says Scott, survived the low temperatures of the Arctic night because "he was with Esquimaux who built him an igloo shelter nightly" (1913, 1:250). Scott does not mention, or had not realized, that Amundsen had shown through his narrative that igloo building is a skill that can be learned. Scott had no trouble convincing himself, however, that new British technology would enable him to "win" the pole, while the "old" technology of Norwegian skiers and Inuit dogsled drivers was difficult (which was certainly true), and therefore unlikely to succeed (far from true, as the tragic outcome was to prove). Scott accidentally got a practical second introduction to skiing during motor sledge tests in Norway prior to the 1911 attempt. Perhaps because Tryggve Gran, going for a spare part for a broken motor sledge, made it look so easy, Scott took him on at once as the expedition's unofficial ski instructor. But Scott's navy mentality was not ready for a new, tricky, and physically demanding skill, certainly not from a foreigner, no matter how many ski championships he had won. Gran's instruction came to be regarded only as a pastime.

Similarly, the experiences of the "sideshow" western expedition to Cape Crozier, which preceded the pole attempt, prompted Scott to muse: "One continues to wonder as to the possibilities of fur clothing as made by the Esquimaux, with a sneaking feeling that it may outclass our more civilized garb—but it would have been quite impossible to obtain such articles" (1913, 1:254). In terms of adaptation to conditions, it was the tight naval canvas and damp-collecting wool that were uncivilized. And, of course, Peary and Amundsen had been obtaining "such articles" since the turn of the century.

The Royal Navy had enjoyed technological superiority—had "ruled the waves"—since Lord Nelson, and change of any sort was seen as a threat to that continued superiority. The consequent tendency to institutional rigidity apparent in Scott's account naturally included the chief weakness of an imperialist power—racism. Scott was able to look on Eskimos as civilized enough to learn from, but not civilized enough to be called civilized. Discussing Leopold M'Clintock's contribution to Arctic travel, for example, he allows that "Esquimaux . . . methods were closely observed and more or less imitated," but continues: "to the English explorers of the nineteenth century belongs the honour of being the first to discover that, again to quote Sir Leopold [M'Clintock]: 'the ice which arrests the progress of the ship forms the highway for the sledge'; they were the first civilised beings to use that highway" (1913, 1:145), as though M'Clintock had discovered sledging instead of merely writing a description of it.

The evidence of Scott's Last Expedition suggests quite clearly that Scott's case was exceptional: few explorers compounded cultural rigidity with command indecision to the same extent. A picture of Scott shows him surrounded by books on the Terra Nova (1:facing 218). Did he read them? We don't know. He may have read the available accounts—his book-lined cabin cannot have been completely for show—but it is certain that a lack of management skills added to institutional, national, and personal prejudices prevented many necessary arrangements from being made. Sixty-four years after Franklin, one could almost say the same of the 1912 Scott party as of the Franklin expedition: although Scott's men hunted a little and canning techniques had become safer than those of Franklin's time, there is still suspicion that their highly processed, "civilized" diet may have been severely vitamin-C deficient, making plausible Huntford's explanation of Petty Officer Evans's collapse (1979, 522–23) and the slow progress of the party on the return from the pole (Scott 1913, 534–35, 581). The other shortcomings of Franklin's earlier Royal Navy expedition—wrong clothing and wrong travel technology—had not really changed at all. Scott's Last Expedition attempts to present Scott and his party as heroic martyrs to the cruel Antarctic, but in fact it shows that they died victims of the cultural rigidity of their own institutions.

Judging from the examples of Franklin and Scott, both characteristic of nineteenth-century British navy polar exploration, it seems that for at least a hundred years after Parry's first contact with the Eskimos, and in spite of an accessible and continuous Scandinavian ski tradition, vital survival and travel information reached those who needed it with remarkably low efficiency. *Scott's Last Expedition* shows how cultural prejudice (a master race mentality based on military and technological superiority, both applicable only to the climates of their origin) prevented adoption of the necessary technology by virtually ignoring scientific narratives that didn't fit the paradigm. In addition to prejudice, the necessity of contact must account for the slowness with which travel and survival information became available. Obviously, practical teaching was available only in the Arctic, and then only in the unlikely context of respect and rapport between Europeans and Eskimos—something else dependent on an appropriately attenuated concept of nation. Meanwhile, Nature, in her polar guise of sparse resources and temperatures capable of freezing exposed flesh in two or three minutes, laughed at the refitted bomb ships, like the *Hecla*, and at such newspaper publishers as Charles F. Hall, that England and America sent to pluck out the heart of her mystery.

Studying the interactions of the natural world with the societies attempting to follow their cultural patterns within it, as seen in the popular scientific narratives of the "heroic age" of polar exploration, is part of the enterprise mapped out by Donna Haraway in her study of the history of primatology: "to envision a different and less hostile order of relationships among people, animals, technologies, and land, . . . to set new terms for the traffic between . . . nature and culture" (1989, 15). Amundsen's *North West Passage* and *South Pole* narratives, in particular, illustrate a "traffic" out of which such an order emerged; dogsled exhibitions have survived even the advent of the airplane to become an "extreme" sport, like the 1988–89 Transantarctica expedition, as well as an occasion for international collaboration, like the 1989 Bering Bridge expedition.

Amundsen's popular scientific narratives exhibit a culture deliberately defining and inadvertently revealing itself at the same time. While they display at least two of the characteristics of imperialistic jingoism (a sentimental *Volksblut* nostalgia akin to Wagner's and Hitler's cult of the Nordic, and a masochistic cult of effort to the point of pain), they seem devoid of racism and crass mercantilism. His narratives recount, sometimes consciously, sometimes unconsciously, the interplay between three cultures: a newly nationalistic type demanding that its patriots make their mark on the world; then, stimulated by the first, a prehistoric indigenous type (common in new nations), in Norway's case having reference to a heritage of hardy navigators and explorers; and, finally, the terrain-constrained Eskimo type with a panoply of survival technologies. It was a blend of the best of all three that was so starkly vindicated in the Norwegian success of 1911–12.

REFERENCES

Amundsen, R. 1908. *The North West Passage: Being the Record of a Voyage of Exploration of the Ship "Gjoea," 1903–1907.* 2 vols. New York: E. P. Dutton.
———. [1912] 1976. *The South Pole: An Account of the Norwegian Antarctic Expedition in the "Fram," 1910–1912.* 2 vols. Trans. A. G. Chater. New York: Barnes and Noble.
Beattie, O., and J. Geiger. 1987. *Frozen in Time: Unlocking the Secrets of the Franklin Expedition.* New York: E. P. Dutton.
Berton, P. 1988. *The Arctic Grail: The Quest for the North West Passage and the North Pole, 1816–1909.* New York: Penguin Books.
Cyriax, R. J. 1963. Arctic Sledge Travelling by Officers of the Royal Navy, 1819–49. *Mariner's Mirror* 49: 127–42.
80 Years Later and a World away, Eskimo Sons of Explorers Meet U.S. Kin. 1987. *New York Times*, 7 June.
Ferris, J. 1989. *Arctic Explorer: The Story of Matthew Henson.* Minneapolis: Carolrhoda Books.
Haraway, D. 1989. *Primate Visions: Gender, Race and Nature in the World of Modern Science.* New York: Routledge.
Huntford, R. 1979. *Scott and Amundsen.* London: Hodder and Stoughton.
———, ed. 1987. *The Amundsen Photographs.* New York: Atlantic Monthly Press.
Lopez, B. 1986. *Arctic Dreams: Imagination and Desire in a Northern Landscape.* New York: Charles Scribner's Sons.
Myers, G. 1990. *Writing Biology: Texts in the Social Construction of Scientific Knowledge.* Madison: Univ. of Wisconsin Press.
Nansen, F. 1890. *The First Crossing of Greenland.* 2 vols. Trans. H. M. Gepp. London: Longmans, Green.
Peary, R. E. 1917. *Secrets of Polar Travel.* New York: Century.
———. [1910] 1986. *The North Pole: Its Discovery in 1909 under the Auspices of the Peary Arctic Club.* New York: Dover.
Scott, R. F. 1905. *The Voyage of the "Discovery."* 2 vols. London: Smith, Elder.
———. 1913. *Scott's Last Expedition.* 2 vols. Ed. L. Huxley. London: Smith, Elder.
Young, W. 1980. On the Debunking of Captain Scott. *Encounter* 44(5): 8–19.

Aggression and Power: The R-complex

and Nuclear Blackmail

ALAN G. WASSERSTEIN

Power is tolerable only on condition that it mask a substantial part of itself. Its success is proportional to its ability to hide its own mechanisms.
—Michel Foucault

The question has been raised, by Michel Foucault and others, whether knowledge is more subversive than we had thought: not a means of liberation from power but a means of domination; not representation of reality but justification of a set of social relations; the proper object not of contemplation but of resistance. Knowledge is said to be in the service of power, whose primacy in human motive we are more inclined to admit than that of intellectual curiosity. *Power* here is intended not in its usual narrow political sense but as a web of social relationships, guaranteeing and supporting life as well as regulating it, enforcing obedience at the top as well as at the bottom of the social system. In Foucault's account power is exerted in clinics, barracks, factories, and prisons; people are willingly counted and classified and, under expert guidance, confess the details of their lives; and these "disciplinary technologies" (Dreyfus and Rabinow 1983) depend on expert knowledge.[1] Power has itself been an object of scientific study not only in the social sciences (where I place Foucault's account of it) but also in the biological sciences, where it goes under the name of aggression. Aggression is one of the names that expert knowledge gives to power; one of the names, that is, that power gives to *itself*. It is, however, a name by which power is not illuminated but disguised. This disguising of modern power is, as Foucault says, the condition of its success. We can observe this dynamic in Arthur Koestler's *Ghost in the Machine*

(1967), Carl Sagan's *Dragons of Eden* (1977), J. D. Bernal's *The World, the Flesh, and the Devil* ([1928] 1969), and Jon Franklin's *Molecules of the Mind* (1987)—texts in which knowledge of aggression sanctions meliorist accounts of moral progress, narratives of liberation through science.

THE CONCEPT OF AGGRESSION

Sigmund Freud and Konrad Lorenz modeled aggression as a potential energy pushing for release and so creating tension.[2] In this model aggression is understood to be a stereotyped and relatively invariant behavior that can be elicited by diverse stimuli. It may be an appropriate response to the environment or a ubiquitous and sometimes maladaptive "discharge." This idea has gained its considerable hearing because it accords with the frequent inappropriateness of human aggression, which amounts, as we are repeatedly made aware, to barbarousness. Although the model has been superseded by some workers in the field, it continues to be the dominant popular conception. Some of its appeal may lie in the idea that brutal behavior is inevitable, therefore excusable. This notion of aggression has accordingly been used to rationalize conservative and laissez-faire politics and winner-take-all capitalist competition; and it can rationalize totalizing and statist politics as well.

These conceptions foster the notion that aggression can be localized to certain brain structures, and experimental data have been gathered to support this idea. A cat in which transection has cut off the midbrain from interaction with neocortical influences responds to minor stimuli with a stereotyped "rage reaction," including "hissing, spitting, growling, piloerection, pupillary dilation, and well-directed biting and clawing" (Ganong 1985, 213). This phenomenon is interpreted as loss of the inhibition of "primitive" by more "advanced" structures; that is, those subsequent in evolutionary development. The same response is elicited by electrical stimulation of areas in the hypothalamus, midbrain, and amygdala. Similarly, the well-known experiments of Jose Delgado, in which a bull was stopped "in full charge" with electrical stimulation, have been taken to support this idea of anatomical localization (Klama 1988, 120). The midbrain, in concert with the limbic system and hypothalamus, is taken to be the anatomical locus of aggression. Again, workers in the field question the relevance of conclusions from brain ablation and electrical stimulation experiments (Klama 1988, 121–23)—the rage reaction is neither the same as nor coextensive with aggression—but the idea that aggression has a localized neural substrate continues to have a wide following.

Paul MacLean (1982) has suggested that the brain has a triune organization. Old structures have been retained through evolution (an inherently conservative process), and new structures have been superimposed. The oldest part of the brain, the midbrain and associated areas, is structurally similar to the brain of reptiles. MacLean supposes that these structures subserve the most basic instincts, including aggression, as well as ritual and stereotyped behaviors. The second tier of brain structure is the old mammalian cortex, including the limbic system, thought to subserve the rich emotional life of mammals, including the raising of young. The neocortex, the third tier, is a development characteristic of higher primates and humans. It subserves language, reasoning, and judgment. MacLean terms the midbrain the "R-complex" (R standing for "reptilian"). Although the R-complex subserves a variety of functions, I use this term to describe a neuroanatomically localized, behaviorally discrete capacity for aggression in people and higher animals.

A familiar argument holds that this discrete aggressive capacity is atavistic, adaptive in the recent evolutionary past but increasingly maladaptive as human cultural progress has outpaced the rate of evolutionary change. The argument is found, for example, in Charles Darwin's private writings. Of these instincts he wrote that "with lesser intellect they might be necessary and no doubt were preservative, and are now, like all other structures slowly vanishing. . . . Our descent, then, is the origin of our evil passions!! The Devil under form of Baboon is our grandfather!" (cited in Klama 1988, 21). The argument for atavism has been advanced in our time by Robert Ardrey, Arthur Koestler, Carl Sagan, Paul MacLean, and E. O. Wilson, among others. Koestler puts it this way: "By all indications the trouble started with the sudden mushrooming of the neocortex at a rate 'unprecedented in evolutionary history' . . . the lines of communication between the very old and the brand-new structures were not developed sufficiently to guarantee their harmonious interplay, the hierarchic co-ordination of instinct and intelligence" (1967, 308). Since the "mushrooming of the neocortex" subserves language and culture, the disparity Koestler describes is between reason and "the body." Koestler does not choose between these venerable antagonists—there is merit on both sides of *that* argument—but he does argue that what is missing is harmony between the two.

Frequently, aggression is linked to another, related instinct, tribal loyalty and the exclusion of the Other, which is said to account for nationalism, war, racism, and religious conflict. Koestler again:

> It was not individual aggression which got out of hand, but devotion to the narrow social group with which the individual identified himself to the hostile exclusion of all other groups. . . . To put it in a different way: to man, *intra-specific differences have become more vital than intra-specific affinities;* and the inhibitions which in other animals prevent intra-specific killing, work only within the group. (308)

Koestler observes that the human capacity for language shares responsibility for what he calls dangerous aggression. That capacity includes the making of abstractions that separate one group from another; it is only after that distinction is made that dangerous aggression occurs. The argument is not that the primitive formations are responsible for aggression, but that they are not in harmony with the advanced structures. The latter, in turn, subserve practices that become dangerous when linked with the energy of the primitive.

As we have pictured it thus far, aggression is a transhistorical *essence*, an explanatory principle that stands outside, and has been constant throughout, our history and had real existence prior to our language and categories. As essence, aggression, whether expressed, sublimated, or repressed, accounts for the way things are. It is the instinct that underlies Darwin's "struggle for existence." Indeed, this is one of the appeals of the notion of aggression: it is continuous and unchanging from biological through cultural evolution. The same processes of natural selection and struggle for existence that drive biological evolution also drive the growth of nations and corporations. This metaphor makes culture, and history, continuous with nature, and so integrates humanity into nature. Power, in Foucault's conception, seems not at all to be the same thing. It exists only in social relations. It is an entirely "superficial" entity that does not account for the way things are but is manifested by them. Nor is power regarded as an expression or sublimation of aggression. But, as we shall see, to regard it in this way is a strategy for advancing power while seeming to curtail it. This is why power has nothing to do with violence. Power depends on authority and on the consent of those it constrains; power ceases to be when it must apply force to get its way. Power depends, in fact, on the authority of knowledge.

THE R-COMPLEX AND RELIGION

The model of the R-complex, or "beast within," is reminiscent of the religious myth of the soul and the passions. The religious myth not only helped to create the idea of the R-complex but has lent it a tremendous, largely unconscious, cultural prestige. Rational scientific consciousness is the soul in modern dress. The aggressive or tribal instincts are the passions, brute nature. Salvation depends on the victory of the soul over the body, or nature. The R-complex is like original sin: we are born into it and cannot overcome it without intercession, religious or scientific, as the case may be. Thus, both science and Christianity sanction what Foucault calls a pastoral power: it "cannot be exercised without knowing the inside of people's minds, without exploring their souls, without making them reveal their innermost secrets" (1983, 214). This is why he likens the practices of psychoanalysis to those of confession (1990). Both the religious and the secular myths justify the control of nature, including human nature, but hide their power under the banner, common to both, of the truth.

The myth of the R-complex shares with religion a vision of paradise regained, where the passions (or the R-complex) are mastered. This sense is nicely captured by Carl Sagan's dragons of Eden, which represent the reptilian R-complex. Dragons are encountered mainly in myths and dreams, which may be functions of the R-complex (used as a metonymy for the irrational in general). The dragons (the R-complex) stand between us and the paradisiacal future to which we might be guided by scientific reason. They guard the return to Eden, to undivided nature, like the Archangel Michael waving his sword. The dragons symbolize also those other reptiles, the dinosaurs, once "rulers" of the earth (like ourselves), who became extinct because of some failure to adapt (Sagan wrote before the advent of the theory of meteor impact, which has diminished our sense of how much the dinosaurs were "responsible" for their own extinction). The implication is that we must "adapt" to the R-complex, using scientific intelligence, or face a similar end—nuclear holocaust brought on by an R-complex out of place in a technical world of enormous material power.

The utopia imagined by scientific optimists like Sagan is one of material abundance. But it also has a spiritual dimension, and in this regard draws strongly on the preceding religious and philosophical tradition. The demise of aggression and tribalism is supposed to eliminate the barriers between people, ushering in an age of what can only be called love; and love is symbolized by a syncitium of personality that might have a kind of immortality. Here is Desmond Bernal, the British crystallographer and polymath, on the subject of a compound brain, or "multiple individual," that scientific progress might bring about in the far future:

the multiple individual would be, barring cataclysmic accidents, immortal, the older components as they died being replaced by newer ones without losing the continuity of the self, the memories and feelings of the older members transferring themselves almost completely to the common stock before its death. And if this seems only a way of cheating death, we must realize that the individual brain will feel itself part of the whole in a way that completely transcends the devotion of the most fanatical adherent of a religious sect. . . . It would be a state of ecstasy in the literal sense. . . . Whatever the intensity of our feeling, however much we may strive to reach beyond ourselves or into another's mind, we are always barred by the limitation of our individuality. Here at least those barriers would be down: feeling would truly communicate itself, memories would be held in common, and yet in all this, identity and continuity of individual development would not be lost. (1969, 43–44)

This "multiple individual" is the material apotheosis of universal reason: reason is made physical and physically shared by the members of the population; the divisions of the body, the separations enforced by aggression and tribalism, are bridged literally. This notion of the Merge is a feature of messianic scientific optimism: in this conception knowledge of nature tends toward an ultimate unity with nature, a sharing in its continuity and "immortality"; scientific reason is the form in which the knower and the known merge. The basis of this hope is the Enlightenment view that nature is manifest in universal laws, and that apprehension of these laws universalizes mind and diffuses it into nature. These ideas are consistent with any totalizing ideology: universal Christian love or (as Bernal himself preferred) the classless society.

The idea of universal nature (and universal knowledge of nature) conflicts with another view of nature, one in which natural competition and diversity, tribalism, and evolutionary branchings are emphasized. A way out of this dilemma is to redefine the tendencies of biological evolution: to say that under its appearance of competition and struggle for existence, and under evolutionary diversity itself, lies a reality of symbiosis on the widest scale; of cooperation between competing elements to form larger and more complex wholes, as when protozoans merge to form multicellular organisms. In this model, life has an inherent tendency to organize itself into ever-larger wholes. Like natural selection, this principle (which is less well established but slowly gaining mainstream adherents) also connects biological evolution with our recent history. Competition and aggression are illusions under which life does its true work of increasing unity. Thus, in the work of Lewis Thomas, individual human beings are seen to be forming collectives to which each relates as do the cells in their own bodies; only these organizations are mental—informational—rather than physical (Wasserstein 1989). Human life continues the progress of nature on the higher level of mind. Bernal's vision is to make these associations literal and physical. In either case the origin of these notions is agape, Christian love. From the viewpoint of the growth of power the outcome is similar. One may hold that nature is separatist and that reason exists to overcome its separations, or that nature is unifying beneath an illusion of separation. In either case, the result is to justify the totalizing of society in the name of forms of knowledge that free us from our fallen estate.

Scientific reason that has overcome the passions or the R-complex is thus imagined to be unconditioned, because conditioning forces—the R-complex, sexual drives, economic forces (manifestations of material want, which would not exist), and the systems of power—would be eliminated. The description of the brain as conditioned is prologue to a narrative of liberation. But what if science itself is one of the forms of power? Then "liberation" would further entangle the individual in power relations. This dynamic is apparent when we consider another similarity between religion and science: the reinscription of apocalypse as nuclear holocaust.

NUCLEAR BLACKMAIL

The control of the beastlike in human nature is sometimes said to be a matter of species survival. Here is a characteristic sentence from Carl Sagan: "There is today in the West (but not in the East) a resurgent interest in vague, anecdotal and often demonstrably erroneous doctrines that, if true, would betoken at least a more interesting universe, but that, if false, imply an intellectual carelessness, an absence of toughmindedness, and a diversion of energies not very promising for our survival" (1977, 237). He goes on to enumerate astrology, flying saucer accounts, modern prophecy, and other efflorescences of popular antirationalist belief. Sagan as scientist shows a bit of siege mentality here: dissent from the rule of scientific reason, even on this small scale, risks apocalypse, the destruction of the species. But he has a genuine fear of the threat of nuclear holocaust, as does Arthur Koestler:

> Thus I am writing this in the year 22 p.H.—post-Hiroshima. For there can be little doubt that in that year a new era started. The human race is facing a challenge unprecedented in its history—which can only be met by taking action of an equally unprecedented nature. . . .
>
> We can only hope to survive as a species by developing techniques which supplant biological evolution. (1967, 323)

This is what I call nuclear blackmail: the stakes are so high, and the threat of extinction so enormous, that no manipulation of human nature to lessen that threat is out of the question. The threat of species extinction by nuclear war justifies radical defensive maneuvers, including manipulations of human behavior that may constitute, in their own right, a kind of extinction of the human. This rationale for the aggrandizement of power over the individual is analogous to the loss of civil liberties in times of national crisis, as with McCarthyism in the 1950s or the "drug war" now.

Sagan argues that learning to handle nuclear power is one of the tests that any advanced civilization—not just our own—must pass; if there are other civilizations in the universe, they must have passed it or they would not have survived (1980, 250). This argument projects the human situation into universal terms. Knowledge of nuclear *physics* is held to be universal, not specific to our forms of power; but the development of nuclear *weapons* is held to be an accident of history, of military necessity, of culture. The social consequences of secrecy and control that nuclear materials require are also "local" and cultural. This line of reasoning exonerates the knowledge while condemning the power; claims knowledge to be neutral and innocent while condemning its social arrangements. But these arrangements are both the mechanisms by which knowledge is accumulated and the consequences of that knowledge.

The nuclear situation is in the highest degree ironic. It is the extreme case of the tendency of science and technology to exert control over the world. But expert knowledge divests itself of responsibility for this power; it claims that the self is divided and projects power into the irrational, primitive, and aggressive parts of the brain. The irrational part of the self (the R-complex) is taken to be responsible for the misapplications of power, while reason is taken to be innocent and even liberating. Science is alleged to be value free, ethically neutral; knowledge of the physical world is supposed to be innocent. Hence the problems with nuclear power lie not so much in science and technology as in other aspects of human nature—the R-complex. The "primitive" human emotional equipment is not adequate to this "advanced" kind of knowledge. We are not *good enough* to handle the power that our knowledge gives us. In this case, then, the science that created this violence has a warrant to make us good enough, to control human nature.

Nuclear blackmail is a special case of what Jacques Ellul (1964) identifies as a general feature of Technology: it is always creating new problems, even with its so-called successes. The needle and syringe facilitate the spread of AIDS; the control of epidemic infectious disease contributes to the population problem; television fosters semi-illiteracy. In the case of nuclear blackmail, the "technical" problem that has been created by nuclear technology is human nature itself. We human beings will have to change in order to make the world safe for our technology. That Technology demands a change of human nature is not new; only the urgency of the demand is said to be greater than ever. Ellul's Technology resembles Foucault's power, not least in the way that power is self-perpetuating, taking off from its last accomplishment to justify a further control.[3]

Power masks itself, projects into the R-complex its impulsion to manipulate, master, and totalize. To power, the R-complex is an aspect of nature; one might even say, of external nature. It is a beast within, a mechanism, a thing—and therefore suitable for manipulation. It does not offend us to imagine such control. On the contrary, control of the R-complex is welcomed as a form of liberation from the bondage of the "instincts." If we project power into the R-complex rather than locate it in our social arrangements, we can imagine freedom as the consequence of those arrangements. They permit us—through science, for example, or through religion—to discover and control that which conditions us. The forms of power that developed nuclear weapons can be projected into the R-complex, which can be dissociated from humanness. We are happy to concede that the bomb was produced at least partly in response to wartime conditions, therefore in response to our inner demons; but we maintain that that condition of the discovery was accidental and unrelated to the knowledge that was developed, and that that condition—the R-complex—can ultimately be done away with. We want to believe that scientific reason pursues a truth that is, or can be, purged of its impulse to power.

The metaphors of the R-complex and of nuclear blackmail are interesting and important in themselves, but they also stand in for basic assumptions about modern life. The R-complex stands in for inherent human resistance to the impositions of technical life. With the R-complex, expert knowledge concedes that individual resistance to technological controls is grounded in human nature. But the concession does not go very far: resistance is characterized as uncivilized, and the R-complex is not a call for "instinctual liberation" but a rationale for more control. Nuclear blackmail stands in, in the strongest terms, for the danger of resistance: it is the extreme case in which the waywardness of the individual impairs productive efficiency; the whole system goes up in flames. Life is so interdependent, we are told, and the technical system so delicately balanced, that wayward actions by individuals could have disastrous consequences for the system, in ways that were not possible before the expansion of technical life. Nuclear blackmail is *the* cautionary tale for our times.

MASTERING THE R-COMPLEX

As to how in the future the R-complex may be mastered, there are several possibilities, including drugs, electrical stimulation, genetic engineering, or something as yet unforeseen. One of the most benign versions of the solution of this problem is, again, to assume that knowledge is not contaminated by power and, indeed, is the liberation from power. Thus, a full scientific description of the R-complex, free of any prescription for overt manipulation, would blunt its power. The act of *recognition* of power, as in psychoanalysis, would neutralize it. It is possible to write about a "nondominating" approach to the R-complex, even if one believes that "survival" is at stake. Thus Carl Sagan believes that "the aperture to a bright future lies almost certainly through the full functioning of the neocortex— reason alloyed with intuition and with limbic and R-complex components, to be sure, but reason nonetheless: a courageous working through of the world as it really is" (1977, 238). He betrays the working scientist's characteristic optimism that the world as it really is can be known through reason. And while reason is primary, he envisions a useful partnership with the irrational part of the brain rather than a transformation of it.

It is comforting to suppose that science will offer only knowledge of the world as it is rather than technical manipulation; that a coming to consciousness, and only that, would neutralize the dragons. But would that be science at all, as we know it? Science is more a matter of manipulation and control than of detached observation, as Jurgen Habermas has suggested:

The idea of a New Science [in which the viewpoint of possible technical control would be replaced by one of preserving, fostering, and releasing the potentialities of nature] will not stand up to logical scrutiny
any more than that of a new Technology, if indeed science is to retain the meaning of modern science inherently oriented to possible
technical control. For this function, as for scientific-technical progress
in general, there is no more "humane" substitute. (Cited in Rouse
1987, 168)

Habermas supposes that the human sciences (*Geisteswissenschaften*) are
inherently different from this description of the natural sciences (*Naturwissenschaften*). But I agree with Joseph Rouse when he argues that the
great difference between them is the degree of experimental manipulation
that is permissible in dealing with human subjects (1987, 198–202); and
I am concerned that those permissible limits will be exceeded under the
influence of nuclear blackmail.

Nor should we take comfort in the idea that aggression is a "mere metaphor," not objective because it is highly conditioned by cultural history and
current societal needs, and therefore of limited use for scientific research.
There is no cause to belittle metaphors; all models of physical reality are
metaphors, analogies between something present (language) and something absent (the real world, especially in its microscopic and subsensory
existence). It is true that the social purposes of the R-complex ought to
make us suspicious of its uses. But that does not in itself imply that the
metaphor must produce bad science. Rouse has argued that science is not a
pure knowledge or theory that subsequently finds technical application. It
is rather a system of manipulation and control performed within the tightly
controlled conditions of the laboratory, in which theory is subsequent to
practice. Theory serves to rationalize and to direct future attempts at experiment, which is no more than a highly developed technique of local
control. Technology extends these means of control to the outside world by
standardizing and normalizing conditions in the world: making the world
as much as possible like the laboratory. It is therefore difficult to believe
that further knowledge of the R-complex, when it comes, will not be a
technical and manipulative knowledge. Though the R-complex may be a
model with more social justification than hard evidence at present, it can
be *made true*. The phenomenon can be created, in Ian Hacking's phrase
(Rouse 1987, 203). That is, the postulate that there is a discrete and ma-
nipulable mechanism, a trigger of aggressive behavior, could yield such a

mechanism or mechanisms. The model of human action implied by this metaphor may be crude and simplistic, but if the more fully developed theory that emerges from our current sense of an R-complex does not do justice to the richness of human behavior, that will not damn the theory. As long as it yields mechanisms of control, the theory will be deemed successful. Swayed by the fact of technical control, people will believe that the theory represents, that it *is,* the real world. But a different set of social relations, and a different sense of what is needed, would give a different knowledge.

Depending on whether science is taken to be liberating or controlling, descriptions of the mastery of the R-complex are benign or paranoid. Koestler imagines a drug to heal the split in the human psyche:

A mental stabilizer would produce neither euphoria, nor sleep, nor mescalin visions, nor cabbage-like equanimity—it would in fact have no noticeably specific effect, except promoting cerebral co-ordination and harmonizing thought and emotion; in other words, restoring the integrity of the split hierarchy. Its use would spread because people like feeling healthy rather than unhealthy in body and mind. It would spread as vaccination has spread, and contraception has spread, not by coercion but by enlightened self-interest. (1967, 337)

This is fantasy not because of the degree of technical facility it assumes, and not because it is reminiscent of the drug culture with which it was contemporary, but because of its benign vision of manipulation: Koestler's "stabilizer" offers freedom from the inner demons without exacting a price in social control. It is a medical solution to a moral dilemma, but strangely innocent of any sense of enhancing the power of doctors. And yet, in one sense Koestler is absolutely on target: people will *want* the power that is applied to them, they will submit to it voluntarily. It will come in the shape of liberating knowledge.

Here, by contrast, is Jon Franklin's paranoid vision of mastering the R-complex. He is considering the future capability of what he calls molecular pharmacology to influence human behavior.

Now, at long last, we will take the final step and assume responsibility for controlling ourselves. That's the sales pitch of the new revolution.

But first, to use the ancient paradigm, there will be the devil to pay. . . . We will have to turn our backs on the duality, and, with it, the faith of our fathers. Molecular psychology represents the most fundamental heresy ever committed by science, and we will have to embrace it.

We will have to look into the mirror, surrender illusion, and make peace with the fact that we're staring at a machine.

We are mechanisms, pure and simple, explainable without resort to the concept of soul. . . .

That is the central, cold, hard, emotionless truth of the revolution in molecular psychology. If we really desire the safety we seek, the safety from chemicals and radiation and war, then we must renounce the romantic, dualistisic view of man. . . .

Molecular psychology demotes our most precious beliefs and ideals to the levels of crude necessity. It is necessary to believe in God, just as it is necessary to have daily bowel movements. Thus it makes a mockery of our religious beliefs, our country, our deep feelings for our wives and our mothers, our antagonisms, our convictions . . . it strips our justifications away and leaves the mechanism stark naked. . . .

The wisdom of molecular psychology is, first, that it allows us to distinguish between [necessary illusions and brutal realities].

And as for the price we pay for that, a certain rule of thumb seems to be emerging. The more important a belief is to you, the more precious an idea, the more pivotal it seems to your existence, the more it seems to bolster the ascendancy of right over wrong, of justice over injustice, the more likely it is to die in the revolution. . . .

Materialism is the soul as well as the body of the future, and those who are psychologically flexible enough to see the new patterns, and put them to use, will shape what's to come. (1987, 259–64)

Here is the familiar use of nuclear blackmail: radical measures are needed if we are to have "the safety we seek." We may have to become—the tired cliché—mere machines. This is tiresome first because of the assumption that science can reduce the world to sheer determinism, helping us to avoid moral choices, when scientific truth is nothing more than the result of choices. It is tiresome also because, preserving a willful naïveté, the author projects into a fearful future what is obviously already the case: life in technical society is mechanized and its freedom circumscribed.

But what is interesting is that Franklin attacks not the R-complex but the neocortex—that is, the field of language, abstraction, and ideology. In effect he agrees with Koestler that abstractions and ideologies are the root of dangerous aggression. Unlike Koestler he proposes to eliminate them altogether, resolving life into mechanistic determinism. This fantasy is absurd, but it is only a version of what is, in fact, a dominant view of human beings in modern society (compare sociobiology and its congeners): they are collections of mechanisms that can be manipulated and controlled. The odd point of Franklin's text is that it takes the R-complex (in its aspect of automatic and stereotyped—that is, machinelike—behavior), once it has been removed from neocortical influences, to be the basis of a viable if distasteful form of human existence. Machinelike human beings would be freed from moral choice and error, nuclear catastrophe included. They would see through their confused motives to underlying neural mechanisms. This would be a liberating expert knowledge; dangerous abstractions and ideologies would lose their hold. But what would liberation mean to beings who were entirely conditioned and without choice?

Whether one establishes control over the R-complex or, like Franklin, simplifies life down to it, the implications for increasing power are similar. Power relations thrive on and depend on the language of essences, of essential human nature—what Richard Rorty calls "metaphysics" (1989). This may be why Foucault and Rorty deny the existence of a category called "essential human nature."

"BRINGING POWER RELATIONS INTO QUESTION"

What impels society in the direction of power, of what Foucault (1990) calls bio-power—power exerted over the human body? The simple answer is desire for greater security from human deprivations: freedom from hunger, cold, sickness, and death. But a moment's thought shows that this is too simple; technical progress has also made life more precarious. Nuclear blackmail (or the population explosion or environmental degradation) is evidence of that. And even if we say that these are unforeseen consequences, condemned in retrospect, of what has always looked like a greater good in prospect, the question remains: Might there not be something dangerous in the accumulation of power in itself? The most relentless of scientific visionaries, J. D. Bernal, foresees this:

The immediate future which is our own desire, we seek; in achieving it we become different; becoming different we desire something new, so there is no staleness except when development itself has stopped. Moreover, development, even in the most refined stages, will always be a very critical process; the dangers to the whole structure of humanity will not decrease, as their wisdom increases, because, knowing more and wanting more they will dare more, and in daring they will risk their own destruction. But this daring, this experimentation, is really the essential quality of life. ([1928] 1969, 68)

Bernal wrote this in 1928, in a late Victorian mode of bloody but unbowed heroism. His view is that knowledge is *dangerous* and the pursuit of it heroic. Along with knowledge comes the risk of self-destruction. One might regard this as an admission of the dangers even of *successful* science: the extinction of the human as we know it, symbolized in the extinction of the self in what he calls the "multiple individual." At any rate, Bernal does not delude himself about the close relation of knowledge and power. He understands that science is in an ironic relation to progress: the changes that it brings about may be opposite to what its practitioners intended. It may threaten their survival in two ways: change them beyond recognition, or destroy them outright.

He resolves this issue into the moral question of courage, one of the ways, from the Christian martyrs to Freud, that knowledge has characterized its liberating struggles against power. He does not identify the scientist with power, the grasping for which would imply not courage but avarice or megalomania. Rather the scientist is like the martyr, reaching for the truth at the cost of life itself. Only there is a curious moral emptiness here, for the truth is not some stable transhistorical principle but a process of development that has no definable or foreseeable goal. This is the irony of progress: we have no idea where it is going or whether we'll like it when we get there. In the end, the justification for this daring is the courage it shows: Bernal's late Victorian views are consonant with William James's ([1902] 1961) famous statement that we regard life as a theater for heroism. The old tradition of heroism in the service of the truth survives, like a phantom limb, when the notion of the truth (as something stable and beyond history) no longer exists. How prescient Bernal was about the dangers of science became evident with the development of the nuclear bomb. And perhaps it is because of the bomb that we do not have his zest for risk taking anymore.

If not for our security, then why the relentless increase of power? Foucault writes as if power were not an explanation at all but a description of the (current) state of things. If we are looking for causal explanations, this leaves his work curiously empty; but he has, as we shall see in a moment, a good reason for reticence. The alternative to his view is to take power as an underlying principle that manifests itself in history, a transhistorical principle. It becomes tempting to associate this principle with an intrinsic biological tendency: aggression. Power in social arrangements could be an expression, a sublimation, of underlying biological aggression—of the R-complex—whose nature is relentless expansion.

This is the dangerous position and one Foucault avoids. For it encourages the hope of doing away with power altogether: of mastering the R-complex, for example. The dream of liberation is to use power (in the form of expert knowledge) to do away with power. But the dream is paradoxical and delusory: power increases with such manipulations, although (and indeed because) people consent to them as liberation. Koestler falls into this trap. Wisely, he notes that it is language (or knowledge), the capacity to make abstractions, that creates the human capacity for war and atrocity. Through abstractions the Other is defined. But then, ignoring his own good counsel, he proceeds with his abstraction of a split in the human brain; he makes the R-complex (or, more accurately, disharmony between neocortex and R-complex) into his Other. He will use all means to overcome it. Underlying this goal is his fundamental and empowering abstraction of human progress. In the name of such language social power aggrandizes itself. *His own* language enforces a power of its own. But he sees in this language not power but knowledge.

Foucault says that "power relations are rooted deep in the social nexus, not reconstituted 'above' society as a supplementary structure whose radical effacement one could perhaps dream of" (1983, 222). I would add that the dream of such effacement is a rationalization, and another means, for the advance of power. Foucault continues:

> A society without power relations can only be an abstraction. Which, be it said in passing, makes all the more politically necessary the analysis of power relations in a given society, their historical formation, their source of strength or fragility, the conditions which are necessary to transform some or to abolish others. For to say that there cannot be a society without power relations is not to say either that those which are established are necessary or, in any case, that power constitutes a fatality at the heart of societies, such that it cannot be undermined. Instead I would say that the analysis, elaboration, and bringing into

question of power relations and the "agonism" between power relations and the intransitivity of freedom is a permanent political task inherent in all social existence. (1983, 222–23)

In this essay my goal has been just such a "bringing into question of power relations." The task is inherently difficult: the strongest power relations are also the most seductive and appealing (that is the source of their strength). Some readers will be put off by the analysis: they will cling to notions of the R-complex and nuclear blackmail as the hope of a scientific society that seems otherwise to have lost its moral bearings. For them, mastering the R-complex is a figurative return to Eden, a contemporary utopian program. Yet those who entertain it take themselves to be not tenderhearted but suitably tough-minded, even courageous. They claim to be willing to face up to great risks: that human nature will be transformed into something unrecognizable, that individual freedom will be abrogated. At bottom they do not take these risks seriously, however, for their utopianism consists in believing that power *can* be deleted from knowledge, and that the result, a world without power relations, literally could not, logically speaking, be oppressive.

But I have raised another possibility: that these dreams built on the R-complex and nuclear blackmail are not blueprints for the future but means of justification, rationalizations, of the power relations that currently define and entangle us. To believe in the R-complex, the beast within, is to believe that human beings need taming, that taming is proper to a technical society, and that such social control and regimentation as now exist are good for us. The practical effect of these beliefs is to ratify and multiply the social arrangements of our technical life under cover of a utopian hope for the future. Universal "remedies" for power, ambitious programs to do away with it, paradoxically return us to a scene of magnified power. It is best to give up on the revolutionary hope of liberation, which might replace one form of control with another more severe. In these circumstances the "intransitivity of freedom" requires not radical and universal remedies but local acts of resistance: calling certain power relations into question. One resists, for example, the temptation to master the R-complex. One resists the very notion of the R-complex and, more generally, the notion of a fixed and essential human nature. The difficulty is in finding ways of thinking of these matters that are convincing, that can overcome the considerable weight these ideas have acquired in our culture.[4]

And what is the status of such resistance, its epistemological or existential status? Does resistance free us of power relations? No. As Foucault suggests, power can be undermined, but there is no society without power relations. We should not imagine an escape from such relations into analysis and critique. A critique that claimed to be a pure knowing, free of power, would be a delusion, not unlike the hope of mastering the R-complex. Calling power relations into question does not abolish them; it merely realigns such relations, and hopes to do so without increasing them. It is wary, above all, of unwittingly fostering the multiplication of power.

NOTES

1. Even where the scientific basis of knowledge is least secure, as in psychiatry, society valorizes the views of experts on which its organization depends. See Dreyfus and Rabinow (1983) for an admirable summary of this position.

2. See Klama (1988, 23–27). I am indebted to this book for an excellent account of contemporary notions of aggression and their scientific shortcomings.

3. There is, however, a profound difference between Foucault and Ellul which should not be ignored. Foucault is at pains to show that power is not an underlying explanatory principle or essence, while Ellul's Technology is just such an essence. Ellul has been criticized for his tendency to treat Technology as a principle independent of history, economics, and politics.

4. Scientific optimists may regard this resistance as a tender-minded failure of nerve before the promise of science. But faith in moral and material progress through science (like belief in an essential human nature) is foundationalist. It is a telos in history that is, finally, an Enlightenment reinscription of the divine. To deny this faith is to regard the world as the product of chance, not will—perhaps the more tough-minded posture after all.

REFERENCES

Bernal, J. D. [1928] 1969. *The World, the Flesh and the Devil: An Enquiry into the Three Enemies of the Rational Soul.* Bloomington: Indiana Univ. Press.

Dreyfus, H., and P. Rabinow. 1983. *Michel Foucault: Beyond Structuralism and Hermeneutics.* Chicago: Univ. of Chicago Press.

Ellul, J. 1964. *The Technological Society.* New York: Random House.

Foucault, M. 1983. The Subject and Power. In H. Dreyfus and P. Rabinow, *Michel Foucault: Beyond Structuralism and Hermeneutics.* Chicago: Univ. of Chicago Press.

————. 1990. *The History of Sexuality*. Vol. 1: *An Introduction*. Trans. R. Hurley. New York: Vintage Books.

Franklin, J. 1987. *Molecules of the Mind: The Brave New Science of Molecular Psychology*. New York: Dell.

Ganong, W. F. 1985. *Review of Medical Physiology*. Norwalk: Appleton Lange.

James, W. [1902] 1961. *The Varieties of Religious Experience*. New York: Macmillan.

Klama, J. [pseud.] 1988. *Aggression: The Myth of the Beast Within*. New York: John Wiley and Sons.

Koestler, A. 1967. *The Ghost in the Machine*. New York: Macmillan.

MacLean, P. D. 1982. On the Origin and Progressive Evolution of the Triune Brain. In *Primate Brain Evolution: Methods and Concepts*. Ed. E. Armstrong and D. Falk. New York: Plenum Press.

Rorty, R. 1989. *Contingency, Irony, and Solidarity*. Cambridge: Cambridge Univ. Press.

Rouse, J. 1987. *Knowledge and Power: Toward a Political Philosophy of Science*. Ithaca: Cornell Univ. Press.

Sagan, C. 1977. *The Dragons of Eden: Speculations on the Evolution of Human Intelligence*. New York: Random House.

————. 1980. *Cosmos*. New York: Ballantine.

Wasserstein, A. G. 1989. Scientific Optimism: A Progress Report. *Raritan Review* 9: 114–33.

Reflective Scientists and the Critique

of Mechanistic Metaphor

MARY ELLEN PITTS

Basic to the tradition of Western scientific thought and deeply embedded in many scientific texts is a series of metaphors, controlling tropes that trail notions of the Newtonian mechanistic universe.[1] The conceptual limitations of these tropes have been examined in recent years by Loren Eiseley, Fritjof Capra, Oliver Sacks, and V. V. Nalimov—all members of the scientific community with outstanding credentials who have turned to writing about science itself and about the role of the individual in relation to science. Their awareness that we participate in shaping the new evolution—cultural evolution—even as our technology can affect physical evolution has led them to speculate on the relation of science and the individual in a world that, since the early twentieth century, has become radically different from the world portrayed in our traditional conceptual metaphors. For these scientists, ours is a world that consists not mechanistically of separate parts to be identified and analyzed, but interactively in a web of ongoing processes.

These four scientists know that in the twentieth century, the universe of quantum mechanics has emerged as vastly different from Isaac Newton's mechanistic universe. In the wake of Werner Karl Heisenberg's uncertainty principle and Niels Bohr's indeterminacy, scientists realize that objectivity is a myth and that gradually, the role of the observer in helping to shape the observed has been extended from physics to the social sciences to medicine. The analytic approach basic to Western science has also come under fire as scientists have begun to realize not only the role of human consciousness in creating what we observe as reality, but the interweaving of all things, effectively exemplified in recent chaos studies in the notion of "sensitive dependence on initial conditions"—the notion that rather than following a linear chain of cause and effect, a storm or a snowflake or an embryo develops through a complex web of determinacy (the genetic code of the embryo, for example) interwoven with indeterminacy (the chance introduction of caffeine or alcohol into the bloodstream). Thus the mechanistic reduction of all things to their component parts has come under attack by scientists, as has the positivist notion that observation of reality can yield full understanding.

Yet this mechanistic attitude toward science is still frequently encountered among an overwhelming number of scientifically nonliterate adults who conceive of science as a sort of religion and scientists as a priestly elite, with the result that they face the danger, as Martin Heidegger argues, not of technology itself but of the "enframing" of technology, which "threatens man with the possibility that it could be denied to him to enter into a more original revealing and hence to experience the call of a more primal truth" (1977, 28). It is this enframing, this epistemological limitation, perpetuated in the public mind through the mechanistic metaphor, that concerns each of the scientists to whom I now turn: Loren Eiseley, whom E. Fred Carlisle credits with having "discovered and developed a new prose idiom for science and literature" (1977, 112); Fritjof Capra, whose texts, though soundly based in science, partake of the Far Eastern flavor of the 1970s; Oliver Sacks, whose alternation between the vast and the specific suggests an oscillation between the metaphoric and the metonymic, between the poetic and the narrative; and V. V. Nalimov, who employs a mathematical metaphor heuristically as he seeks an alternative to mechanistic understandings.

LOREN EISELEY

Loren Eiseley recognizes the function of metaphor in science, maintaining that science could not have reached its present stage without "the hook of analogy, the root metaphor" (1980, 3). Even if the analogy is false (his example is the machine analogy as construed by eighteenth-century thinkers), he notes that it may color the thinking of generations, despite the assumed objectivity of the scientific method (1980, 20). Further, metaphor makes possible the movement of understanding from one specialty to another: "The successful analogy or symbol . . . frequently allows the scientist to leap from a generalization in one field of thought to a triumphant achievement in another" (1978, 274). Such analogies function as do literary ones, "whose meanings similarly can never be totally grasped because of their endless power to ramify in the individual mind" (1978, 274).

But Eiseley also recognizes the freedom of metaphor to evolve in unexpected ways. He argues that a root metaphor, having escaped "from the professional scientist into the public domain . . . may undergo further individual transformation and embellishment," indeed may become "as free to evolve in the mind of the individual as are the creations of art" (1978, 275). The metaphors that both express and contain ideas make science itself subject to human imagination, to "enrichment" that could not occur within the framework of "objective" empiricism. Yet metaphor, like most things for Eiseley, has a double edge; it brings both gain and loss. If it helps to effect the leap from one domain to another, if it brings enrichment that a (hypothetical) nonfigurative language could not bring, it nevertheless can intrude on thinking when least expected.

This is the case with mechanistic metaphors, whose history Eiseley traces from the political and military machine of the Roman Empire (1969, 43) to the Newtonian *Machina coelestis*—according to which, complete in some views with cogs and wheels, "the machine reigned" (1980, 15); to Hutton's world machine, which was "the quintessence of Newtonian world order" (1958, 328); and even to the late nineteenth-century quest for the *Urschleim*, which revealed that, still, "mechanism was the order of the day" (1959, 35). Even among Darwin's contemporaries, after evolutionary theory had reintroduced indeterminacy, Eiseley observes that "the machine that began in the heavens had finally been installed in the human heart and brain" (1980, 55). Today, Eiseley contends, humankind is awed by the machine, and "the march of the machines" is in our blood as we look to space to expand our domain (1970, 70).

Clearly Eiseley engages the machine metaphor and its problematic. Yet the machine itself, the vehicle of the metaphor, is sometimes a source of ambivalence even for Eiseley, who leaves the reader caught in a mystery as he quotes Garet Garret for the epigraph to the central chapter of *The Invisible Pyramid*: "Either the machine has a meaning to life that we have not yet been able to interpret in a rational manner, or it is itself a manifestation of life and therefore mysterious" (1970, 73). Playing on Garret's ambivalence and his own, Eiseley draws the reader into this uncertainty. He speculates that the machine may lead humankind to become a kind of "spore bearer" of the first complex form of life to enter space, or it may lead us to program the human personality into "the deathless machine itself" and, in a sense, to escape mortality (1970, 76–77). Yet his attitude itself marks a recognition of the power of the metaphor.

Further, the actual human encounter with the machine suggests its danger as metaphor. Eiseley argues that the triumph of the machine as a cultural artifact can be nothing more than an "atavistic return to the competition and extermination represented in the old biological evolution of 'parts' "—a particularization rather than a holistic view—unless a "triumph" of understanding accompanies it (1958, 350). Because the machine as metaphor carries with it an assumption of human control, we are in danger, he contends, of seeing the world as "an instrument or a mere source of materials" (1970, 143). And the machine as metaphor suggests collective humankind, "man in the mass marching like the machinery of which he is already a replaceable part" (1971, 175). Functioning collectively and seeing itself as dominant over nature rather than functioning as part of nature, humankind may also see itself as only a projection of the machine, as a tool. Thus the human "grows convinced that he is himself only useful as a tool" (1978, 269).

The machine concept, in addition to developing "collective" humankind and leading us to see ourselves alternately as dominant over nature and as mere projections of the machine, also creates a notion of "a permanent, rather than dynamic, balance" in nature (1980, 73), a balance the Darwinian view cannot sustain. Eiseley's answer to the problematic of the metaphor is a balance of "scientific" and "poetic" knowledge and a reintegration of humankind with nature—a return to sympathy with, participation in, and identification with the natural world.

And so in "The Bird and the Machine," after describing an account of a mechanical mouse that can outperform its live counterpart, Eiseley writes, "It's life I believe in, not machines" (1959, 182). Then, with his characteristic turn to personal narrative, Eiseley tells a simple tale of capturing a hawk in an abandoned house on the western desert. By attacking his captor's hand, the bird enabled his mate to soar through a hole in the roof. The captured hawk rested quietly and hopelessly in his captor's hand, though his eye gave "a fierce, almost indifferent glance" (1959, 189). The next morning the narrator released the bird, which flew straight upward toward his mate still circling above. A mechanical mouse may "perform" in a maze better than a living mouse, but Eiseley thinks of the hawk:

Ah, my mind takes up, on the other hand the machine does not bleed, ache, hang for hours in the empty sky in a torment of hope to learn the fate of another machine, nor does it cry out with joy nor dance in the air with the fierce passion of a bird. Far off, over a distance greater than space, that remote cry from the heart of heaven makes a faint buzzing among my breakfast dishes and passes on and away. (1959, 193)

This single moment of participation in and with life reverberates over the years, and the tale illustrates Eiseley's emphasis on participation in, rather than separation from, the natural world. Thus, to the extent that the mechanistic metaphor that underlies the scientific epistemology leads to separation from nature, Eiseley finds the metaphor dangerous; and in narratives such as this one he describes the human ability to participate in life. The moments he recounts are memorable: releasing the sparrow hawk, wrestling with a young fox cub in "The Innocent Fox" (1969, 194–212), separating a hen pheasant and a large blacksnake in "How Natural Is Natural" (1980, 151–81), joining the mad savior of starfish in "The Star Thrower" (1969, 67–92).

As one of Eiseley's most powerful comments on the detachment and fragmentation of the scientific epistemology, "The Star Thrower" gives us Eiseley as he encounters himself, on a coast "set apart for shipwreck," in the form of a "cold world-shriveling eye" circling in his skull (1969, 72–73). Through the densely textured essay, Eiseley leads the reader from a comparison of landscapes (which, he suggests, shape our outlook) to a recollection of tornadoes (a constant reminder of contingency); from his youth in Nebraska to the ghostly, posturing figure of the trickster (the human symbol of contingency) in a primitive tribe. In the form of an old madman whose self-appointed task is to rescue dying starfish on the beach at Costabel, Eiseley finds "one of the last great rifts in nature" and, recalling his mother's eye from a torn photograph (an eye that bespoke isolation and frustration because of her deafness), expresses his love for "the lost ones, the failures of the world" (1969, 87, 86). The rift, he concludes, is actually "a joining: the expression of love projected beyond the species boundary by a creature born of Darwinian struggle," even though this declaration is "like the renunciation of my scientific heritage" (1969, 87, 86). As he joins the star thrower, he does so with a sense of projection into life, of participation beyond the boundaries that separate species. Eiseley's declaration of affection for the "lost ones" and his symbolic throwing of the "stars" signal his acceptance of an epistemology *other* than that of science, his acceptance of the coexistence—and equal importance—of that knowledge we know as poetic.

FRITJOF CAPRA

Like Eiseley, Fritjof Capra sees an emerging paradigm shift[2] "from the mechanistic to the holistic conception of reality" (1983, 16). He argues that academics are more inclined than most to "subscribe to narrow perceptions of reality" and to the fragmentation that grows out of the mechanistic paradigm (25). Like a growing number of reflective scientists, Capra looks for a shift in perspective, for a way to perceive "dynamic patterns of change" (26). In our "globally interconnected world," he argues, "biological, psychological, social, and environmental phenomena are all interdependent," and the Cartesian worldview is inadequate to describe it (16). In looking at the present paradigm and its basis in the Enlightenment, Capra notes with concern that the scientific method is recognized as "the only valid approach to knowledge," and that its understanding of the universe as composed of building blocks that can be isolated and studied is accepted as the only viable approach to truth (31). Such an outlook, he argues, also sees society in terms of competition and struggle, and of "unlimited material progress to be achieved through economic and technological growth" (31). In contrast, he notes the cyclical nature of the Tao, the reality that is continual flux and (noncausal) change. Ceaseless motion, continual process, flow—all these, in Capra's view, constitute the Chinese philosophy of a natural order of dynamic balance (35). Change, in this view, is not a *result* but an essential part of the "continual cosmic process" of the universe (37).

Capra argues that rational thinking tends to be analytic, linear, and fragmented, whereas the other realm of knowledge, intuitive knowledge, arises from direct experience through expanded awareness and "tends to be synthesizing, holistic, and nonlinear" (38). Painstakingly tracing the course of René Descartes's, Isaac Newton's, and John Locke's philosophies, Capra finds that even when evolution and electrodynamics could not be accounted for within the classical mechanistic scheme, the underlying ideas of classical physics were nonetheless assumed to be correct (74).

To answer the problems inherent in the mechanistic paradigm, Capra calls for a recognition of alternative paradigms and for the reeducation of the general public. Hence, from modern physics, he stresses "words like *organic, holistic,* and *ecological*" used to describe a universe no longer a machine composed of discrete parts but "one indivisible, dynamic whole whose parts are essentially interrelated and can be understood only as patterns of a cosmic process" (78). Capra also attends to Heisenberg's explanation of the uncertainty principle, which calls on the textile metaphors that also recur in contemporary literary theory: "the world thus appears as a complicated tissue of events," Heisenberg writes, "in which connections of different kinds alternate or overlap or combine and thereby determine the texture of the whole" (cited in Capra 1983, 81).

Thus causality is displaced by probability, and separate particles or isolated elements are replaced by "nonlocal connections to the whole" (86). Matter seems to be structured similarly to mind, so Capra turns to David Bohm's work, which uses the hologram (in which a small piece of the image reproduces the whole) as an analogue for the universe and the brain that perceives.[3] Or similarly, in the words of James Jeans, "the universe begins to look more like a great thought than like a great machine" (cited in Capra 1983, 86). Again in a related vein, Capra cites the physicist Geoffrey Chew's explanation of the bootstrap approach, according to which nature is not reducible to fundamental "parts," but all parts are interrelated, the universe is portrayed as a web, and this web is dynamic (92–93). In a sense, "every particle consists of all other particles," for subatomic particles are "interrelated energy patterns in an ongoing dynamic process," and each "involves" the others though none "contains" another (94).

Capra's long chapter on the systems view prepares the reader for the emerging paradigm. The systems approach treats ecology, economy, society, and the human body as interactive systems. Thus we are part of "a living system composed of human beings and social organizations in continual interaction with one another and with the surrounding ecosystems on which our lives depend" (390). Because all these systems are self-organizing, balance and flexibility are likely to "be achieved through self-transcendence—breaking through a state of instability or crisis to new forms of organization" (396). Living systems participate in a "nonlinear interconnectedness" that must be understood in terms of process rather than the stasis implied in the Newtonian paradigm (391, 322).

In arguing that we need to transcend the mechanistic metaphor of the universe, Capra suggests that the notion of "physics as the basis of all science" must be abandoned in favor of "mutually consistent concepts" that may "describe different aspects and levels of reality, without the need to reduce the phenomena of any level to those of another" (97). He believes that the needed paradigm shift will begin with medicine because "the functions of a living organism that do not lend themselves to a reductionist description—those representing the organism's integrative activities and its interaction with the environment—are precisely the functions that are crucial for the organism's health" (104). Despite the success of the mechanistic analogy of a living cell with a factory, Capra senses that the cell is increasingly recognized as an organism itself. Our society is, Capra contends, at a turning point, a point where the old paradigm is bankrupt and the new paradigm is only emerging. The systems paradigm—beginning, Capra suggests, in medicine—will signal the end of the mechanistic hierarchizing of fields of knowledge and the acceptance of disciplinary complementarity.

OLIVER SACKS

Although Capra sees the new paradigm emerging first in medicine, Oliver Sacks finds medicine less promising. A highly literate neuropsychologist, a reader of philosophy, metaphysics, and literature, as well as a highly successful practitioner, Sacks, in the chapter in *Awakenings* (1983) titled "Perspectives," considers the embedding of the mechanistic metaphor in Western thinking, focusing on its effects on the science of neurology. Accepting the effectiveness of mechanistic thinking in systematizing and categorizing, he nevertheless calls it "the madness of the last three centuries" (1983, 205). For Sacks, the mechanistic view makes medicine less than human and provides a reason for his reexploration of the case study as a genre:

> It is this Newtonian-Lockean-Cartesian view—variously paraphrased in medicine, biology, politics, industry, etc.—which reduces men to machines, automata, puppets, dolls, blank tablets, formulae, ciphers, systems, and reflexes. It is this, in particular, which has rendered so much of our recent and current medical literature unfruitful, unreadable, inhuman, and unreal. (1983, 205)

Gottfried Leibniz, he notes, stressed that mechanical understandings make sense only in that they subserve the metaphysical (1983, 205), but we have not followed Leibniz. For Sacks, the mechanistic view has led us toward measurement that is not measurement: witness Rebecca, a retarded woman

Sacks describes in *The Man Who Mistook His Wife for a Hat* (1987). Rebecca failed miserably when tested for skills that required seeing patterns or solving problems, but felt herself held together by the associative, nonlinear understanding of words and poetry, and by narrative and theater. Unable to function on the conceptual level, Rebecca was renewed, almost reconstituted, in a verbal world:

> She needed the world re-presented to her in verbal images, in language, and seemed to have little difficulty following the metaphors and symbols of even quite deep poems. . . . [S]he was at home with poetic language. . . . Metaphors, figures of speech, rather striking similitudes, would come naturally to her, though unpredictably, as sudden poetic ejaculations or allusions. (1987, 179)

Seeing Rebecca "recomposed" in the presence of "a narrative (or dramatic) mode," Sacks writes of a "feeling of two wholly different modes of thought, or of organisation, or of being. The first schematic—pattern-seeing, problem-solving—this is what had been tested, and where she had been found so defective. . . . But the tests had given no inkling of anything *but* the deficits, of anything, so to speak, *beyond* her deficits" (1987, 181). Measurement, Sacks contends, is a scientific given, an accepted tool, and a useful code, but it is only a tool. The problem with many of the evaluations that our mechanistic science has devised, Sacks argues in language reminiscent of Eiseley's, is that "they only show us deficits, they do not show us powers; they only show us puzzles and schemata, when we need to see music, narrative, play, a being conducting itself spontaneously in its own natural way" (1987, 181). In terms of narrative, Rebecca was complete; "measured," she was a human failure, a misfit. Seeing Rebecca in terms of two "modes—so damaged and incorrigible in the one, so full of promise and potential in the other," Sacks sees promise and potential in all patients (1987, 182). Through Rebecca and others he shows that medical science has "paid far too much attention to the defects of our patients" (1987, 183).

For Sacks, then, one answer to the problems that the mechanistic paradigm brings is to reexamine the "seeing," the observation, of science. He quotes Ludwig Wittgenstein: "The aspects of things that are most important for us are hidden because of their simplicity and familiarity" (cited in Sacks 1987, 42). Sacks's task is to focus on such observation, to apply Wittgenstein's epistemological statement to both physiology and psychology, to situate the observation in relation to the human being as well as to the scientific institution. Modern medicine, Sacks writes in *Awakenings*, is far too concerned with "subjects" that could as well be rats as humans; it "dismisses our existence, either reducing us to identical replicas reacting to fixed 'stimuli' in equally fixed ways, or seeing our diseases as purely *alien* and bad," requiring attack with the weapons of a whole arsenal of war metaphors as well as technology (1983, 205). Such notions, rooted in the language of objectivity, "increasingly dominate the entire landscape of medicine"; they "are as mystical and Manichean as they are mechanical and inhuman, and are the more pernicious because they are not explicitly realized, declared and avowed" (1983, 205–6). Thus Sacks attempts to dis-place the mechanistic and battle-laden language of modern medicine with alternative metaphors, returning to Louis Pasteur and Claude Bernard for the notion that the pathogen is less important than the terrain, and even to Michael Faraday's "History of a Candle" to emphasize the "history" and the "life" of everything concrete (1983, 206).

Sacks's reevaluation of "seeing," closely related to the "objective" lan-guage of science, leads to another answer to misunderstandings rooted in the mechanistic paradigm: recognition of an alternative to scientific under-standing, with its emphasis on fragmentation and measurement—recogni-tion of a mode of knowing through language, even through what he calls "metaphysics." In the medical literature on parkinsonism and L-DOPA, for example, he finds " 'facts,' figures, lists, schedules, inventories, calcula-tions, ratings, quotients, indices, statistics, formulae, graphs, and whatnot; everything 'calculated, cast-up, balanced, and proved' in a manner which would have delighted the heart of Thomas Gradgrind" (1983, 207). Yet *"nowhere,"* he continues, "does one find any colour, reality, or warmth; nowhere any residue of the living experience; nowhere any impression or picture of what it *feels* like to have Parkinsonism, to receive L-DOPA, and to be totally transformed" (1983, 207). Medicine, he argues, needs "an approach, a language . . . that must be both particular and general, com-bining reference to the patient and *his* nature, and to the world and *its* nature" (1983, 208–9). The terms that medicine needs are "at once per-sonal and universal, concrete and metaphorical, simple and deep," for they "are the terms of *metaphysics,* or colloquial speech" (1983, 209). These terms are *meta*-physical in what Sacks calls "the simplest and deepest terms we know," the terms of "health" and "disease"; they are literally, in the sense of the Greek roots, "the things after the physics" (1983, 209). Sacks argues for the need to explore these simple terms, avoiding "superficial definitions and dichotomies, and to *feel* (beyond the range of formulations) the intimate, essential nature of each" (1983, 209).

One means of exploring this *meta*-physical language is through the ordinary-language paths of narrative and drama, "literary" genres. Medical science, he contends, has been concerned "far too little with 'narratology,' the neglected and needed science of the concrete" (1987, 183). Sacks considers himself "a naturalist and a physician both," "equally interested in diseases and people," and "equally, if inadequately, a theorist and dramatist," "equally drawn to the scientific and the romantic" (1987, vii). (The term *romantic* is one that Sacks appropriates from his correspondence with the Russian neurologist A. R. Luria, who practiced "a warm science, a 'Romantic Science,' . . . heretical, secret and loving" [1984, 210].) It is patients, the spokespersons for the human condition, who lead him to reflection: "Constantly my patients drive me to question, and constantly my questions drive me to patients" (1987, vii). It is patients who bring the dramatic, the unexpected, the "romantic" to be understood. Hippocrates introduced the case history, the notion of the "course" of disease. But Sacks's goal is more. He seeks "to restore the human subject at the centre—the suffering, afflicted, fighting, human subject"; case histories, he contends, must be "deepened" "to a narrative or tale" that shows the person in relation to the disease (1987, viii). Sacks compares his patients to the archetypal figures of ancient fables because they live in worlds where "the scientific and the romantic . . . cry out to come together . . . at the intersection of fact and fable" (1987, ix).

Such poetic understandings—for Sacks, as for a number of writers trained as scientists but confronting writing as the source of an understanding *other* than that of science—provide an organization, an opposite of analysis, division, and compartmentalization, which gives the greater understanding. And Sacks moves from natural language to another language, the felt language of music. In *A Leg to Stand On* he describes his own battle with neurological damage from a hiking accident; the experience led him to read the early neurologist Henry Head, who describes the "kinetic melody" of actions. Reading Head, Sacks finds "two modes of thought, two realms" joined: "On the one hand, there was the language of analysis, science, and neurology: 'sequence,' 'series,' 'complex procedures.' . . . On the other hand, Head spoke of 'completeness,' 'wholeness,' 'perfection,' and 'melody,' the language of the visionary, the poet" (1984, 200). And he remembers his father's description of Head: "He was the most rigorous of scientists, but he was a poet too. He *felt* the music of movement and speech, but as a neurologist he could not *explain* it" (1984, 200). The last two chapters of *A Leg to Stand On* thus chronicle Sacks's search for the point where neurology went wrong so that he can help it find wholeness again. In the writings of Civil War neurologist Weir Mitchell he finds an openness to "the richness of the phenomenal in health and disease" (1984, 206). Yet "by the 1880s, with the rise of scientific neurology and its new and powerful mechanical concepts of 'function,' the possibilities of such naturalism were decisively closed" in favor of a " 'puppetology'—a purely mechanical vision of man and creature as reflex puppet" (1984, 206). At the end of his quest, Sacks finds that "neuropsychology had crashed in ruins about me" and "empirical science had failed me as a guide to the musicality of action and life" (1984, 217).

Sacks remembers his early encounters with Parkinson patients whose actions seemed mechanical and recalls that he asked W. H. Auden what was missing in them. Auden's instant response was, "The music of motion," to which he added Novalis's phrase: "Every disease is a musical problem, every cure a musical solution" (1984, 219). In the shared rhythm of music and motion and language he finds a centering point, valorized afterward in his reading of Emmanuel Kant. His love for narrative, for poetry, for the language of feeling, for the "romantic" in science is valorized in the awareness that "there need not be a gulf, or a split, between joyous life and grey theory—but, on the contrary, a wonderful coming together" (1984, 221). The language echoes Eiseley's at the time of his epiphany when he chose to join the star thrower: the rift in nature "was not a rift but a joining" (1969, 87).

In these "joinings" of understandings, of scientific and poetic knowledge—shared in even the language that Eiseley and Sacks use, and in the complementarity that Capra espouses—there is movement away from the particularized, fragmented world that the mechanistic paradigm has given us. Martin Heidegger's words reverberate: "There was a time when it was not technology alone that bore the name *techne*. . . . Once there was a time when the bringing-forth of the true into the beautiful was called *techne*. And the *poiesis* of the fine arts also was called *techne*" (1977, 34). Heidegger's answer to the "danger" of technology, the answer he calls the "saving power," is art, but only if reflection on art remains open to questioning, not closed by the enframing of technology.

V. V. NALIMOV

The reflective scientists whose work I have considered clearly attempt to suggest alternatives to the mechanistic model. The last writer I treat, the Russian biologist, statistician, and philosopher V. V. Nalimov, also exemplifies the Heideggerian openness in his questioning of art, science, and knowledge itself. Unlike Eiseley, Capra, and Sacks, however, he must be viewed in the context of literary theory in recent decades—specifically in terms of the concept of textuality—and in the context of concepts that migrate not from the sciences into the humanities but from the humanities into the sciences. Nalimov's quest for an alternative to the mechanistic view of the world is unusual because he reflects on the "humanization" of science, incorporating both a metaphor from mathematics and a concept from literary theory. He not only undermines the dominant mechanistic paradigm and argues for a systems view, but he uses a mathematical model as a heuristic metaphor, drawing examples from multiple disciplines and bringing them together through the notion of textuality.

In his critique of science, Nalimov finds Kurt Gödel's demonstration of the limits of deduction "the most powerful epistemological result ever obtained" (1981a, 206). In addition to this epistemological realization, Nalimov finds in recent philosophical investigations of scientific methodology both a "humanization of knowledge," so that "concepts from the humanities penetrate the foundation of all scientific activities," and a "'cybernetization' of science" (1981a, 207). These two effects on science form, as it were, two poles of the corrective he proposes for the mechanistic paradigm. Instead of suggesting an alternative metaphor such as a web, a tissue, or a hologram, Nalimov draws on the "cybernetization" of science as he uses the Bayesian theorem heuristically in discussing language, evolution, and even science itself.[4] Yet he also draws on the humanization of knowledge to search for the "lost unity" or "indivisibility" of knowledge itself, through his repeated exploration of the textual metaphor that migrates from linguistic and literary studies to become the focus of his quest for understanding the diverse forms of human knowledge.

Nalimov takes all knowledge as his province. He argues that belief in the "omnipotent formalism of logic" (1982b, 49) and reliance on the machine metaphor have shaped scientific thinking since the eighteenth century. Such "omnipotent formalism," he argues, is embedded in the Newtonian mechanistic view of the world and in the thermodynamic analogues that emerged in the nineteenth century. In an attempt to move away from the mechanistic view and its accompanying faith in the formalism of logic, Nalimov turns to the "text" and the Bayesian heuristic, through which he attempts to display "the unity of the World" as expressed "in the *language* of its texts" (1982b, 1). He argues that the textual paradigm offers a means of rethinking the world, of subverting the mechanical paradigm. In the uncertainty and nonlocality of quantum physics, in transpersonal trends in psychology, and in existential philosophy Nalimov finds evidence of a new paradigm overlapping the old. This new paradigm, he argues, remains rough and unfinished, but it threatens to displace positivism, determinism, behaviorism, and fragmentation, all of which were based on "the simplicity and mechanisticity of nature and the omnipotence of its laws" (1982b, 49). In *Faces of Science* he tells how his book on language (originally titled *Probabilistic Model of Language* and published in Moscow in 1974, translated and published in the United States in 1981 as *In the Labyrinths of Language: A Mathematician's Journey*) was originally published "only because all its constructions were based (though heuristically) on a mathematical model"; when the book was expanded, an extended discussion ensued over whether the Soviet humanities board or the technology board should publish it (1981a, 212). Yet knowledge, Nalimov argues, is broader and more human than the current fragmentation would indicate. "It is high time," he concludes, "to stop looking at the world with photoelements, thermoelements, and other measuring devices and begin to acknowledge our right to look at the world of those who manipulate these devices and interpret the data they produce" (1981a, 212). Nalimov's work can be seen as an attempt to think through the "rough" parts of the new paradigm that he sees emerging from two paradoxically linked sources: the humanities and cybernetics.

Basic to the paradigm that Nalimov explores is the notion that the world itself is a text through which "we begin to comprehend the metaphor by Heidegger-Ricoeur: '*Man is a language*'" (1982b, 1). Nalimov does not repeat the seventeenth- and eighteenth-century notion of the Book of Nature. His view is complex, employing the concepts of number and rhythm as basic to the textual structure of the world. The underlying factors in translations of the world as text (these translations are all fields of knowledge) are rhythm and number. For Nalimov, the individual "texts of the physical World are revealed to us as a numerical arrangement of things in time and space," just as "a crystal is rhythm imprinted in a stone" (1982b, 1). He notes, in Heideggerian fashion, that the privileged notions of *word* and *speech* trail meanings that include *counting, calculus, number, group,* and *category* (1982b, 6). Number, he argues, may function "as a basic category of consciousness"; thus Nalimov asserts that we need to examine the surprising similarities of the languages of the world text if we are to comprehend the generalized text (1982b, 12, 22).

Nalimov, then, contends that all human interactions with the world have in common the use of a language, whether natural language or the language of symbols, mathematics, or music. In each of his four books translated into English he explores a ramification of language. *In the Labyrinths of Language: A Mathematician's Journey* (1974, trans. 1981) in effect affirms the linguistic *episteme*. "The study of language," he writes, "is a study of thinking" (1981b, xv). In this book he attempts to focus on the "frontier" where language, cognition, and consciousness meet. In *Faces of Science* (1981) he addresses science as a reflection of culture and the role of language in creating ideas, with emphasis on the twentieth-century movement from determinism to probabilistics as a basis for understanding natural and human phenomena. In *Realms of the Unconscious: The Enchanted Frontier* (1982), Nalimov takes his probabilistic approach to the unconscious, exploring randomness as a carrier of knowledge rather than of ignorance. In this daring exploration he approaches medieval mysticism and Eastern and Oriental meditation as sources of understanding, and he reports on experiments he and his colleagues performed using a diverse group of subjects (engineers, pilots, painters, and some mental patients) who used group meditation techniques to explore the semantic fields of the words *freedom, slavery,* and *dignity* (1982a, 185–86). And in *Space, Time, and Life: The Probabilistic Pathways of Evolution* (1985), he uses the Bayesian theorem to explore the world itself as a text, positing a "semantic universe" in which meaning is probabilistically "unpacked," equating the spontaneous emergence of information with the human faculty for self-transcendence.

In Nalimov's terms, then, the world is a text, though in his system the overarching text parallels the Saussurean *langue* (Nalimov carefully documents his reading of Saussure). This *langue*/text is a system structured as language is structured, with individual utterances that constitute "translations" of the world text into "human languages comprehensible to us"—that is, in the form of biological or physical knowledge or of the poetic "interaction of the personal rhythm with the rhythm of the world" (1982b, 1). Nalimov emphasizes the need to distinguish between discrete and semantic components of the *langue* that he posits, for he finds in "any sufficiently rich language . . . two elements: the discrete (the word or any of its discrete analogs) and the contiguous (the continuous semantic field underlying words)" (1982b, 26). At this point, a definition of the concept of *field* in its scientific sense is in order. The concept of field suggests, according to N. Katherine Hayles,

> that reality consists not of discrete objects located in space but rather of a underlying field whose interactions *produce* both objects and space. It further implie[s] . . . that there is no exterior, objective viewpoint from which to observe, for one is always already within the field, caught in and constituted through the very interactions that one is trying to describe. Field models were associated with the emergence of inherent limits on what can be articulated. (1990, xi–xii)

Nalimov's discrete and contiguous elements echo Roman Jakobson's metaphoric and metonymic poles[5] and lend support to Umberto Eco's contention that "metaphor can be traced back to a subjacent chain of metonymic connections which constitute the framework of the code and upon which is based the constitution of any semantic field" (1984, 68). The discrete in biology, to look at one language, is the individual, species, or superspecies, whereas the continuous is "the entire possible variety of morphophysiological properties ordered on a numerical continuum," or "biological semantics" (1982b, 26). For Nalimov, the "semantic" field—whether in biology, chemistry, or physics—is the element of contiguity, of continuous possibility. This view emphasizes evolution as a means of conserving characteristics as well as of transforming them; evolutionary potentiality functions parallel to the notion of *langue,* the system of possible utterances. The point is the balancing of stability versus spontaneity and possibility— a point to which Nalimov, like Eiseley, turns emphatically when he considers the interplay of stability (the need of the species to remain stable) and potentiality, an interplay that chaos science treats in terms of a system that is both deterministic and free.

As his new systems paradigm overlaps the old mechanistic paradigm, Nalimov does not urge abandoning logic or rejecting traditional Aristotelian logic as hopelessly outmoded, but embracing an expanded logic of fuzzy concepts—a probabilistic logic that comprises *prediction,* not *determination,* of thought, word, or action in the world text. In using the language of probabilistic logic, Nalimov attempts to expand the use of logic to include conclusions drawn "from fuzzy, probabilistically weighted concepts," to "juxtapose freely the entire range of knowledge accumulated by mankind," whether from ancient mathematicians or medieval mystics or transpersonal experiences, and to compare the incomparable on the basis of his conviction that the human unconscious is the reservoir of potentiality in human understanding (1982a, xiv–xv).

Within the text as the global metaphor for the world and the probabilistic view of the semantic field, chance becomes a rich source of understanding. In the semantic field of the world text in all areas of knowledge, Nalimov finds chance "rehabilitated": whereas it was once "a synonym for nonsense," it has become productive (1982a, 29). In fact, "chance now expresses knowledge rather than ignorance" (1982a, 14). In probability theory, a random sequence of symbols represents "maximum complexity" rather than ignorance (1982a, 14). Among human beings, Nalimov contends, "any action is preceded by a *decision*," and this process is "always trying; it always leads to bifurcation" (1982a, 17). Chance, then, in its complexity, generates information and leads to an unpredictable splitting of directions, or bifurcation.

Further pursuit of this concept requires consideration of some insights of information science and chaos studies. Information, of course, is separated from knowledge in Nalimov's schema. As in information science, *information* is equated with complexity, with data—bits or bytes—so that greater complexity means greater possibilities of the new, greater creativity. As Hayles notes, when Claude Shannon merged information and entropy (defined thermodynamically in terms of increasing molecular randomness, or the tendency toward disorder—to use a word loaded with cultural baggage, "chaos"), he led to conceiving of chaos as the source of the new; this conception of randomness as maximum information brought a new perspective for understanding chaos.[6] The new awareness, Nalimov contends, took more than twenty centuries to emerge (1982a, 14). Nalimov's probabilistic view of the world text embraces the view that chaos—randomness or disorder—is both rich in and generative of information. His attention to scale parallels yet another contribution of the broad area loosely called chaos studies, in which a view of the whole or a mapping of the process of the whole demonstrates the sensitive dependence on initial conditions that makes it possible for the motion of a fish to create an eddy or of a butterfly's wings to influence weather.[7] Randomness, then, is creative as pattern affects pattern, part disproportionately affects whole. Applying the textual metaphor, with its insights from information science, to evolution, Nalimov finds this translation of the world text comprehensible "only by means of a multiplicative model that enables us to see the change of the existing pattern as a whole affected by another pattern" (1982a, 27). Indeed, he argues that science itself is "a self-organized system which behaves as a living organism" and evolves as do organisms (1981b, xvii).

The role of chance in Nalimov's world text thus leads to a strong sense of the creativity that occurs on the microlevel and influences the macro. He argues that "new texts are always a result of free creativity realized on a probabilistic set which may be regarded as an unexposed semantic universe or *nothing*, the semantic vacuum or, metaphorically, an analogue of the physical vacuum" (1982a, 29). The physical vacuum from which he analogizes is not emptiness as we know it, but a state in which physical processes manifest physical properties; like the term *chaos*, the notion of a *vacuum* trails cultural meanings and has been modified as scientific understandings have changed.[8] Once thought of as totally empty space, the vacuum state is the state of lowest energy, but an electromagnetic (photon) field does not cease to exist when there are no photons, even though it is no longer observable. When enough energy exists in the photon vacuum, a photon appears and the vacuum state becomes observable. The point is, for Nalimov, that "contemporary physics has come close to the concept of *non-existence* (the unobservable state) as a potential basis for *reality* (the observable state)." Because "the charged point modifies space around itself," evolution of the field involves simultaneous change "in an infinite number of points" (1982a, 39). Thus, by defining nonexistence and reality in terms of the unobservable and the observable, Nalimov brings into consideration the previously marginalized concepts of *non* or *nothing*, which emerge not as absence but as the realm of potentiality.

For Nalimov, then, the notion of *chance* is epistemological rather than ontological (1982a, 41). Chance and the fuzziness of concepts are to be explored rather than suppressed in reading the world as text (1982a, 42). The world is seen not in terms of sharply differentiated objects but in terms of probabilistically fuzzy descriptions, and nonexistence must be considered even as existence is considered (1982a, 42, 73). Nonexistence is, Nalimov contends, "the ultimate reality of the world" (1982a, 73). If consciousness itself is, as Nalimov argues, a geometrical representation of the world (1982a, 71), and, ultimately, geometry is Nothing (not in the ancient Eastern sense of nothingness, or nirvana, but in the sense of a source of potentiality, of the potential text [1982a, 79]) then Nothing is the semantic vacuum from which information emerges. As this semantic field evolves, changes occur in infinite points; thus evolution suggests both the stability of the *langue* and the creativity of the *parole,* whether in biology or in everyday language.

What is unique in the animate world, however, is that it contains "elements of the Future in the Present" (1982b, 47). When evolution is viewed on the global scale, we find "*spontaneous* emergence of new information"—an "unpacking" of the potentiality of nature, an unpacking that is not tied to cause and effect but occurs probabilistically and spontaneously (1982b, 49). In the world of animate life, Nalimov equates the spontaneous emergence of information with "the faculty of the system for self-transcendence" (1982b, 49), which can occur in the conscious realm as humankind changes its physical, social, and cultural structures, as well as its own consciousness. He parallels this self-transcendence with the way a language can be used to discuss its own structure (1982b, 50). The self-transcendence of language is metalanguage; the metalanguage of humankind is evidenced in our ability to talk about both present and future. For Nalimov, then, the element of spontaneity and creativity associated with language is present in the world as text, and though he warns of the dangers of the term *creative,* he emphasizes the linguistic analogue in describing the creativity of evolution (1982b, 53).

In this probabilistic world of creative evolution, space and time are also defined in terms of the text. Space, because of its receptiveness, becomes "a *potential* text, its receptacle (such that it is related to the text by its filling)" (1982b, 69), and time is "the grammar of texts of the World" (1982b, 58). This notion draws on the ancient image of the word as a receptacle of thought, of the brain as a receptacle for knowledge. In this textual analogy, space and time function as Jakobson's metaphoric and metonymic poles, invoking contiguity and similarity, metonymy and metaphor. Because of its relationship of "filling," space is the contiguous No-thing, the metonymic potential from which all narratives derive; and time, as "grammar," is the structure that enables a given "language" to make distinctions of similarity/dissimilarity. The individual *parole* is generated from the semantic vacuum of Nothing. Texts result from the "comprehension by the subject of the mute meanings of the space" (1982b, 69). Thus the reader also creates the text. Within this system, which may be applied to genetics or to written texts, the principle is "*the unity of the whole and freedom of its constituents*" (1982b, 55). The whole is that which unfolds or is "unpacked," while the constituents vary fractally; as in ordinary language, the exact meaning can be predicted but not precisely determined. Although change occurs spontaneously, paradoxically it is neither entirely free from determinism nor entirely predictable.

The implications of Nalimov's linguistic analogizing, his world as text, are significant. His system demonstrates a blending of speculations from physics and poetry, from biology and philosophy. He approaches randomness as it has evolved in information science. Not only does he criticize the widespread mechanistic model, he attempts to replace it with a heuristic metaphor from nonlinear mathematics. Nalimov's notion of the text is not identical to, but clearly is influenced by, notions of the text as described by Saussure and recent literary theorists. In the twentieth century we have seen migrations of notions from physics to literature, from evolution to economic theory. As Hayles argues, these migrations result not from "influence" but from culturally conditioned concerns of thinkers in a variety of disciplines (1990, 4). Various interpretations of indeterminacy have emerged in the physical sciences and have either migrated to or evolved at almost the same time in literary studies and epistemology. Nalimov's contribution to what Hayles calls "isomorphisms" is his heuristic use of statistics to develop a metaphor for the whole, his absorption of indeterminacy into a structure that *is* a whole but allows freedom to its constituents. The notion is far from a return to free will; it is a heuristic metaphor derived from the probabilistic studies that have made information science a model for human thought. Nalimov's model involves the individual—evolutionary creature or linguistic utterance—in a complex web of unity and randomness, of constraints and freedom, and he draws this model from a remotivated understanding of the randomness that in this century has seemed to spell the world's undoing.

But Nalimov's textual paradigm is important, too, in its overt reversal of the pattern usually perceived for migration of concepts: here, instead of migration of a model from the (privileged) sciences into literary studies, Nalimov deliberately asserts the impact of the humanities on the sciences, and he uses the textual—the literary—paradigm to explain scientific understandings. In the context of the textuality of the world, of both science and literature, Nalimov's paradigm occurs on three levels (as does my reading of the shift from the mechanical to the textual paradigm): (1) reading scientific writing as we read literature, for Nalimov reads Darwin, Heisenberg, Wittgenstein, and Hermann Hesse in terms of the "languages" they represent; (2) reading science itself as an institution, a semiotic text to be examined in terms of the analogy to Saussurean linguistics; and (3) exploring the rhetoric of the text through probabilistic explorations of statistics and evolution, rhythm in music, and structure in language in an effort to displace the traditionally embedded mechanistic image of the world and its inhabitants and to visualize the world as a textual web of determinacies and indeterminacies.

These scientists—Eiseley, Capra, and Sacks exploring the route to interdisciplinarity through popular understanding, and Nalimov exploring interdisciplinary understandings through a mathematical heuristic and a literary metaphor—join others in attempting to displace the mechanistic metaphor as a shaper of Western thinking. Why, one wonders, should highly qualified writers devote such intensity to a simple metaphor? As science has turned away from rigid determinism, positivism, particularism, and mechanism, the public seems nevertheless to have clung to science as authority. Where modern scientific theories lead to complementarity and indeterminacy, to interaction of the observer with the observed, the popular hierarchizing of authority over nonauthority undermines the systems or interactive view necessary to understanding. Recognizing this undermining, these writers turn to language for understandings that can be valued equally with mathematical language and technological measurement—for an understanding of feeling in Eiseley's expression of love across the boundaries of species or time or social convention, for an understanding of the organism's need for interaction with its environment in Capra's quest for complementary levels of reality, for an understanding of the joyousness of human movement through rhythm and music in Sacks's quest for a "neurology of the soul," and for an understanding of knowledge itself in terms of other systems of knowledge in Nalimov's quest for a heuristic of probability.

NOTES

1. For a discussion of the essential metaphoricity of language and the notion of conceptual metaphors, see Lakoff and Johnson (1980).

2. Capra and Nalimov both refer to Thomas S. Kuhn's notion of the "paradigm shift" (1970). In his postscript to *The Structure of Scientific Revolutions*, Kuhn explains two basic senses of the term *paradigm:* "the entire constellation of beliefs, values, techniques, and so on shared by the members of a given [scientific] community," and "the concrete puzzle-solutions which, employed as models or examples," form the basis for continued puzzle solving (175). Science, he continues, consists of communities, and the members of a given community share both practices and

commitments. Thus he defines *paradigm* in terms of a "disciplinary matrix," which suggests a commonality of both understandings and practices (182).

3. For further readings on the "holographic paradigm" as a model for the brain's functioning, see Wilber (1982). Drawing on evidence that storage of information in the brain is not limited to specific segments but takes place throughout the brain, Bohm and Karl Pribram have concluded that, in an oversimplified but useful paraphrase, "the brain is a hologram, interpreting a holographic universe" (Ferguson 1982, 22).

4. Concerned with the essential metaphoricity of language, Nalimov overtly employs the Bayesian formula as a metaphor, paralleling his use with the "qualitative theory of differential equations which allows one to obtain information on the behavior of the solution of a system of equations without solving them and, consequently, without knowing the numerical values of parameters" (1982a, 42n). He stresses the metaphorical (hence polysemic) and the multiplicative (rather than additive) aspects of this formula, which he also treats as a syllogism. In his metaphorical application of this formula from statistics,

$$p(\mu/y) = kp(\mu)p(y/\mu),$$

μ represents the semantic field of a word (a continuous scale) and p indicates probability. In his application, $p(\mu)$ represents the "tendency for the word meaning to be realized" (1981b, 87); it is "the likelihood function which defines the distribution of the semantic content of sentence y, provided our attention is drawn to the meaning of word μ" (1981a, 80). He emphasizes that $p(\mu)$ is a "fuzzily given" premise (1982a, 42), as is $p(y/\mu)$, which is a conditional distribution function signifying "the *spontaneity* of choice . . . that exists only as an unrealized potentiality" (1982b, 28). From these two premises is derived the corollary $p(\mu/y)$, which is "the a posteriori probability that defines the distribution of the meaning of word μ in sentence y" (1981a, 80); k is a normalizing constant (1982b, 28). Nalimov emphasizes the interaction of the discrete and the continuous, which "deepens our comprehension both of the functioning of human everyday language and of consciousness itself" (1982b, 26).

5. Roman Jakobson links metaphor to romanticism and symbolism, metonymy to realism; further, he links metaphor to poetry and metonymy to prose (1971, 1113–16).

6. Hayles (1990, 50–51). For a complex discussion of the conflation of the term for randomness in thermodynamics with information theory and the cultural understandings possibly involved in this conflation, see ibid., 31–60.

7. For a journalistic general introduction to chaos studies, see James Gleick's informative and popular *Chaos: Making a New Science* (1987). The articles by Saperstein (1984) and May (1976) provide further insights into chaos studies.

8. Hayles traces the cultural meanings attached to the term *chaos* from its classical sense of the substance from which the world was formed to the post-Renaissance sense of chaos as the opposite of order, through the reinforcement of the conflict

of order and chaos with the emergence of thermodynamics, to the contemporary reevaluation of the relationship between chaos and order (19–23).

REFERENCES

Capra, F. 1983. *The Turning Point: Science, Society, and the Rising Culture.* New York: Bantam Books.

Carlisle, E. F. 1977. The Poetic Achievement of Loren Eiseley. *Prairie Schooner* 51: 111 29.

Eco, U. 1984. The Semantics of Metaphor. In *The Role of the Reader: Explorations in the Semiotics of Texts.* Bloomington: Midland–Indiana Univ. Press.

Eiseley, L. 1958. *Darwin's Century.* New York: Anchor-Doubleday.

———. 1959. *The Immense Journey.* New York: Vintage–Random House.

———. 1969. *The Unexpected Universe.* New York: Harvest–Harcourt Brace Jovanovich.

———. 1970. *The Invisible Pyramid.* New York: Charles Scribner's Sons.

———. 1971. *The Night Country.* New York: Charles Scribner's Sons.

———. 1978. *The Star Thrower.* Intro. W. H. Auden. New York: Harvest–Harcourt Brace Jovanovich.

———. 1980. *The Firmament of Time.* New York: Atheneum.

Ferguson, M. 1982. Karl Pribram's Changing Reality. In *The Holographic Paradigm and Other Paradoxes: Exploring the Leading Edge of Science.* Ed. K. Wilber. Boulder: Shambhala.

Gleick, J. 1987. *Chaos: Making a New Science.* New York: Viking Press.

Hayles, N. K. 1990. *Chaos Bound: Orderly Disorder in Contemporary Literature and Science.* Ithaca: Cornell Univ. Press.

Heidegger, M. 1977. The Question Concerning Technology. In *The Question Concerning Technology and Other Essays.* Trans. W. Lovitt. New York: Harper-Colophon.

Jakobson, R. 1971. The Metaphoric and Metonymic Poles. In *Critical Theory since Plato.* Ed. H. Adams. New York: Harcourt Brace Jovanovich.

Kuhn, T. S. 1970. *The Structure of Scientific Revolutions.* 2d ed. Chicago: Univ. of Chicago Press.

Lakoff, G., and M. Johnson. 1980. *Metaphors We Live By.* Chicago: Univ. of Chicago Press.

May, R. M. 1976. Simple Mathematical Models with Very Complicated Dynamics. *Nature* 261: 459–67.

Nalimov, V. V. 1981a. *Faces of Science.* Ed. R. G. Colodny. Philadelphia: Institute for Scientific Information Press.

———. [1974] 1981b. *In the Labyrinths of Language: A Mathematician's Journey.* Ed. R. G. Colodny. Philadelphia: Institute for Scientific Information Press.

————. 1982a. *Realms of the Unconscious: The Enchanted Frontier*. Ed. R. G. Colodny. Trans. A. V. Yarkho. Philadelphia: Institute for Scientific Information Press.

————. 1982b. *Space, Time, and Life: The Probabilistic Pathways of Evolution*. Ed. R. G. Colodny. Trans. A. V. Yarkho. Philadelphia: Institute for Scientific Information Press.

Sacks, O. [1973] 1983. *Awakenings*. New York: E. P. Dutton.

————. 1984. *A Leg to Stand On*. New York: Summit.

————. 1987. *The Man Who Mistook His Wife for a Hat and Other Clinical Tales*. New York: Perennial-Harper.

Saperstein, A. M. 1984. Chaos—A Model for the Outbreak of War. *Nature* 309: 303–5.

Wilber, K., ed. 1982. *The Holographic Paradigm and Other Paradoxes: Exploring the Leading Edge of Science*. Boulder: Shambhala.

Contemporary Ecophilosophy in David

Quammen's Popular Natural Histories

ALLISON BULSTERBAUM WALLACE

The poet says the proper study of mankind is man. I say, study to forget all that; take wider views of the universe. . . . When another poet says the world is too much with us, he means, of course, that man is too much with us. . . . Man is but the place where I stand, and the prospect hence is infinite.
—Henry David Thoreau

One provocative coincidence in the history of American ideas was the mid-nineteenth-century rise of a new approach to natural history—an integrative one with a wide field of vision, dubbed "ecology" by 1866— at about the same time that Henry David Thoreau's holistic views of man and nature were maturing in books that would have far-reaching influence. It would be difficult to claim that early ecological thought included any trace of Thoreau's feeling that the human species is "but" one of many, for nineteenth-century ecology did not often question values, certainly not human value itself. But this hint of antihomocentrism in Thoreau has become a major dimension of recent ecophilosophy, itself the child of a century of ecological field research. Contemporary ecology continues to build for us a construct of nature rarely seen prior to Thoreau, his mentor Ralph Waldo Emerson, and their contemporary Charles Darwin (all contributors, in their way, to the new paradigm). The resulting ontological lessons have been translated into moral precepts informing many of the books and magazines currently at the front of a long tradition of popular American writing on natural history: the *hows* and *whys* of nonhuman nature are no longer the only—or even the primary—subjects explored.

Since Thoreau's day both the science of ecology and the ecological habit of mind have taken root in Western culture, although their attendant view of humankind as part of the biotic community, not a species removed from or superior to it, does not rule our collective attitude toward nature. True, an environmental movement distinct from the earlier conservation movement has been afoot since the early 1960s; a Green party has begun to attract modest attention in United States politics, having already made its presence felt in several European countries; recent apprehension about polluted waterways and air, holes in the ozone layer, overburdened landfills, and the too-slow grind of bureaucratic machinery charged with cleaning up manmade wastelands has made *ecology, environment, biodegradable,* and *recycle* the buzzwords for the 1990s; and courts are hearing more and more cases debating the protection or restoration of natural areas. Still, a genuinely ecological, ecocentric worldview, one that routinely takes nonhuman entities into its field of vision and perhaps into its kinship or ethical circle, remains a rare find in an American, albeit less rare than it once was.

Whereas evolutionary thought began in the nineteenth century to examine connections among and within species over vast stretches of time, the sister science ecology took for its object contemporaneous contexts in nature: interdependent, not necessarily benign, relationships among species and habitats, eventually called biotic communities. This emphasis on contexts has contributed to a gradual shift from the old metaphor for earthly life as a great "chain" of being toward the less hierarchical picture of a "web" or circle of being. Encompassing other creatures along with humans and other landscapes along with those bearing the shadow of a human figure, ecology (literally, "house study") dislodges humanity from its traditional seat of land-lordliness. Taken to its most recent ecophilosophical conclusions, this approach to nature almost invariably affirms what Thoreau suspected: man is indeed too much with us, too much on our minds—too much even now the measure of all things, centuries after Protagoras gave us the phrase.

When contemporary ecophilosophy challenges the fundamentally homocentric Western mindset, it challenges the most pervasive, most long-lived concept of Western modernity—humanism, both Christian and secular. This potentially confronts literary naturists working from the ecological perspective with a dilemma, since their very vocation as writers may well stem from deep involvement with human life and meaning, that is, from their own humanism. Moreover, their literary training is likely indebted as much or more to works about men and women as to those about snakes or mayflies. But their work in the life sciences constantly reminds them that the large-brained ape who began to walk upright was a contingent, not an inevitable, fact of history—and of very recent history at that. So when the dust settles after such a writer's struggles between homocentric allegiance to humanity (indeed, his or her own life), on the one hand, and to eco-centrism, on the other,[1] it appears that the real target is neither humanism nor humanity at large. Rather, it is human arrogance: anthropocentrism, homocentrism, lately also called "speciesism." Against claims that *Homo sapiens* represents the pinnacle of creation, that "humankind is what nature has been trying, all these millennia, 'to be'" (Turner 1990, 39), ecocentrists pose an attitude felt to be closer to the "truth" of nature: nature "does not exist for us, had no idea we were coming, and doesn't give a damn about us" (Gould 1990, 30), no matter how much we care about ourselves.

Read the post-Thoreauvian, post-Darwinian American nature writers in chronological order: they increasingly address the problem of anthropo-centrism in our thinking about and our treatment of nature. At worst, they assert, such egoism results in wholesale destruction of species and habitats, possibly including our own (although even this much anthropocentrism in arguments urging human restraint is suspect in some circles). At the very least, homocentrism consigns us to life in a hall of mirrors, no doubt full of wonder at all things human, but even so, an existence less large, less inter-esting, and less interdependent (again, the lesson of ecology) than it could be. Bridging the "two cultures" of science and the arts, John Burroughs, John Muir, Mary Austin, Joseph Wood Krutch, Henry Beston, Aldo Leo-pold, Sally Carrighar, Rachel Carson, Edward Abbey, Peter Matthiessen, Annie Dillard, Edward Hoagland, Barry Lopez, John Hay, Diane Acker-man, David Rains Wallace, Gary Nabhan, and many others throughout the twentieth century frequently share the thinking of those working more in philosophy than natural history per se. Variously called ecocentrism or biocentrism, this mindset implies a reconsideration of value distinctions traditionally made between the human and the nonhuman.

Neither full-time novelists or poets nor full-time scientists, these part-timers rove both fields, but the two professional cultures have yet to fore-ground nature writing in their canons—which situation may be just fine by the writers. This fact is very much to the point I wish to make about the influence of ecology on them personally as well as on the genre in which they work. From the writers' view, interdisciplinary work serves as a corrective to the tunnel vision resulting from professional specialization. Here we come back to Thoreau's wider view, for many thinkers a more interesting, and perhaps more urgently needed, macroscopic vision which allows, even demands, the integration or synthesis of many different kinds of information.

One such literary naturist is David Quammen, author of *Natural Acts* (1985) and *Flight of the Iguana* (1988), who addresses the difference between himself and science writers like Lewis Thomas, Freeman Dyson, or Stephen Jay Gould: "What I am is a dilettante and a haunter of libraries and a snoop. The sort of person who has his nose in the way constantly during other people's field trips, asking too many foolish questions and occasionally scribbling notes" (1985, xiv). Unlike the others, he claims not to "do science," but this seems a smokescreen to draw attention away from the extensive primary and secondary research that goes into his essays; his two books include long lists of "partial sources." Yet although Quammen is more a man of letters than of science—he wrote a master's thesis on Faulkner and has published a handful of his own novels and stories—his growing reputation rests on his "Natural Acts" column in *Outside* magazine and his two ensuing essay collections. Of the literary academy's scant attention to such works, by Quammen or others of his ilk, Thomas Lyon asserts, "the humble standing of nature writing is simply the corollary of our culture's preference for literature that focuses on humans and their personal and social lives" (1989, xiv).

In place of such anthropocentrism, Quammen takes a witty, ironic view of human uniqueness: "True, we have a fancy hand with an opposable thumb. True, we have an elaborate brain capable of memory, foresight, iambic pentameter, and malice. . . . But not nearly so true," he adds, "is the anthropocentric presumption that *Homo sapiens* represents some sort of evolutionary culmination, embodying all the latest and best ideas. We are no such thing." This "humbling reality" he then illustrates by comparing the human animal with "a plump, homely gob of living matter known as the sea cucumber," to which I shall return. All of this appears on the first page of the first essay of his first collection, *Natural Acts*. The lead essay in *Flight of the Iguana* takes matters further by asking, "How should a human behave toward the members of other living species?" In six pages he poses the question three times.

Quammen introduces *Natural Acts* with the warning—or lure— that his tastes run toward "good questions" rather than "good answers" and "toward certain creatures that are conventionally judged repulsive, certain places that are conventionally judged desolate, certain people and ideas that are conventionally judged crazy" (1985, xi, xv). Hence the book's subtitle, *A Sidelong View of Science and Nature.* The opening remarks in *Flight of the Iguana* put the case in different, and stronger, terms: despite our probable first impressions of the bizarre creatures he will describe in the following pages, Quammen tells us, "nature is *not* a freak show," and his is "not a book full of geeks."

> On the contrary, the point here is simply nature itself on a good day. On a *normal* day. . . . These unpopular beasts I seem to have gathered here . . . are the natural and true-born practitioners of life on this planet, legitimate scions of organic evolution, as surely as are the white-tail deer or the parakeet or the puppy. If we ourselves can fathom them only in the context of carnival canvas . . . the problem is probably our own. (1988, ix–x)

The reference to familiar animals, the phrase "legitimate scions" to describe unfamiliar ones, and the comment about "our problem" are all telling. The hallmark of Quammen's work is a sustained effort to overcome readers' typical xenophobia, as he calls it, toward bats, octopuses, anacondas, spoonworms, scorpions, bamboo, Venus flytraps, and the like. Straight facts and exploded myths are the first order of business, then, but not at the expense of the wondrous. He shares Diane Ackerman's belief that fear or "indifference" toward nonhuman nature "is always based on ignorance" but can "be replaced by fascination, which is based on knowledge" (Ackerman 1990, 77). Quammen likewise finds up-to-date scientific information his chief tool for heightening readers' awe and respect for nature's "Eloquent Practices [and] Natural Acts" (title of part 4 in *Natural Acts*).

Of his two primary subjects, nature's "surprising" facts and "human attitudes toward those intricacies," Quammen says they are "all ineluctably connected"—ineluctably ecological (1988, xi). Grounded in this science of natural relationships and participating in contemporary ecophilosophy's argument for kinship between human and nonhuman beings, *Natural Acts* and *Flight* clearly aim to revise the long-cherished view that man is the measure of all things—if by "measure" we typically mean an aesthetic or moral valorization of the human at the cost of all other natural entities. Indeed, Quammen would improve upon the great chain of being by replacing the angels as humankind's link to divinity with wild geese, "the images of humanity's own highest self" insofar as they "live by the same principles [liberty, grace, devotion] that we, too often, only espouse" (1988, 234).

My analysis of Quammen's revisionist project begins, then, with several techniques he uses singly or in combination to chip away at our exalted view of ourselves: exposing our sins against the planet's ecosystems; pointing out where anthropocentrism has led to genetic degradation—if not outright extinction—of certain species; occasionally attributing "low" animal traits to the human being, while at other times endowing animals with some of "our" best qualities (as in that example of the wild geese); and defending creatures used or abused for frivolous or otherwise ignoble purposes. All this may sound like one great misanthropic enterprise, but it is not, or at least not wholly so. As I shall demonstrate, Quammen's work chiefly insists on the *qualitative connections* to be found between humans and the Other, to the ultimate improvement of existence for all concerned.

But first, some examples of the bad news according to Quammen. Between paragraphs describing the delicate, unique ecosystem found in New Mexico's Tularosa Basin, he points out that the first nuclear bomb ("Trinity") was detonated nearby and that the contemporary White Sands Missile Range located in the area is today "America's largest land-area shooting gallery" (1985, 199). An essay about a quirky river snorkler gives half its attention to the "awesome environmental pillage" inflicted upon a Montana river by a copper smelter, such that a snorkler emerging from it nowadays will "glow in the dark" (1985, 94–96). Elsewhere Quammen describes how human meddling with fish populations turned another river into a "grayling ghetto" (1985, 108). If the mosquito has come to be (for humans) "The World's Most Despised Animal" (1985, 24), it may be the result of "a strong case of anthropocentric press-agentry" (1985, 27). Quammen sees this "devil" as "one of the great ecological heroes of planet Earth" (1985, 27) to the extent that it aids in rain forest preservation by making such places "virtually uninhabitable" for people (1985, 27). He describes the havoc currently being wreaked in tropical forests—with verbs like "mowed," "pulped," "corded," and "gobbled"—a "catastrophe" which nothing has helped to avert before now except the two thousand species of mosquitoes living there (1985, 28). And essays about the evolution of the common dog and the uncommmon cheetah detail these creatures' genetic degradation over the centuries since people first took an interest in them, (no) thanks to "implacable, greedy human impulse[s]" (1988, 154).

The more forcefully Quammen wishes to speak, the more his criticism takes on misanthropic hues. In *Flight*'s title essay on the Galápagos, for example, he explains the fragility of island ecology in the face of any change that comes too quickly, especially that brought on by "whalers and other forms of sea-borne human predator," a phrase he uses now and then to emphasize "natural" predation carried too far (1988, 174). And in each book Quammen compares human beings to rabid dogs, insofar as both, unlike a Gila monster or a black widow, creatures we typically fear, are known to kill "without good reason." (This last comes as an oblique attack, in the section of *Flight* entitled "The Moral Ecology of a Desert," on armed border patrolmen and Central American death squads. Crossing the Sonoran hills, Quammen and two members of the southwestern Sanctuary movement are afraid of the former, whereas the family of Salvadoran refugees they are trying to help fear both. Of such people and also the Gila monster Quammen studies to screen his real investigation, he repeatedly says, "There are tricks of perspective involved" [1988, 179].)

Quammen gives both implicit and explicit attention in *Natural Acts* to one major source of the inhumanity toward nonhumans illustrated above: sometimes called the doctrine of final cause, Quammen names it "the argument from utility" (1985, 112). He notes Western culture's tacit translation of Jeremy Bentham's ideas into a prescription reading "the greatest good for the greatest number"—"of *humans*" (1985, 115). Traditionally the theologic has been that all of God's creations, animate or otherwise, have as their "final cause" or raison d'être the service of God's favorite creation, the human family. Obviously this is one point on which theology and secular practice have usually been in happy agreement, resulting in widespread exploitation of natural entities for several centuries and then a slower, "managed" program of exploitation called conservation.

Sometimes Quammen appears merely to poke fun at the utility notion. "Any creature so grotesquely improbable as the common bat must lend itself to some grotesquely improbable human use," he writes of a failed attempt to use bats as World War II firebombers. "The idea was to refrigerate these bats into hibernation, see, fit each with a small payload of napalm and a little bitty parachute, see, drop thousands like that from planes over Japanese cities, and hope for the best." But one feels the smile fade as Quammen adds, "No I'm not making this up. Your government. The research cost two million dollars." It didn't work, he says, because the parachutes were just a "little too bitty" (1985, 8–9, 12).

More often, practices stemming from the final cause mindset do not lend themselves to much humor. As Quammen describes the similar shape and structure of an octopus eye and a human eye, in order to illustrate the concept of convergent evolution, he tells of Pacific Coast scuba divers wrestling octopuses to shore for sport. After these "championships," the octopuses "were carefully released back into the sea. No harm done. . . . Right?"

Except that the octopus is actually "very fragile . . . highly developed with a very sensitive nervous system"—meaning that rough handling can send one into shock and premature death. So Quammen imagines the octopus's eerily familiar (after all, its eye is like ours), "troubled gaze" as a "desperate" attempt to "communicate"; in effect saying, "*Listen. We know who you are. And we've seen what you do*" (1985, 22–23).

Perhaps because the idea has been so long taken for granted that nature's beings exist for the gratification of human pleasure and need, even environmentalists use it to defend endangered species or ecosystems. They do so despite the fact that it means losing the war in order to win the battle: to argue for protection of an entity on the grounds that we might one day find it useful (nutritionally, medicinally, minerally, recreationally, whatever) may mean that someone will one day find a way to use it.[2] "The logic is as solid as it is dangerous," Quammen warns, discussing the debate that raged over whether to halt construction of a major dam in order to save the last snail darters on Earth: "The best reason for saving the snail darter was this: precisely because it is flat useless. That's what makes it special. It wasn't put there, in the Little Tennessee River; it has no ironclad reason for being there; it is simply there" (1985, 112).

In the same essay he says that another fish, the grayling, is not good for any *use* but is nevertheless "good for a person." Not always consistent, the writer who decries wrestling octopuses for sport is himself an avid freshwater sportfisherman. But of these fish he says, "You catch grayling to visit them: to hold one carefully in the water, hook freed, dorsal flaring, and gape at the colors, and then watch as it dashes away" (1986, 116). To ice his case for the purposelessness of yoking utility and wildlife, Quammen deftly works in the point that "the Earth is generally warming; it is in fact falling inexorably into the sun, and the sun itself is meanwhile dying. So all wildlife on the planet is doomed to eventual elimination, and mankind is only et cetera" (1985, 117). Bringing a species to its end prematurely would, whatever the short-term gains, do no one any good down the road. Of course, this is a double-edged sword; ever since the conservation movement got under way, many people in the business of searching out the planet's resources, such as geologist Charles Park, have felt we should make the "best" of what there is while we still have it (McPhee 1971, 22). But Quammen and his contingent would rather rage against the dying of the light. One thinks here of environmentalists' short-term "victory" in late 1990 on behalf of the spotted owl's old-growth forest habitat in the

Pacific Northwest, despite the human impracticality of preserving all that timber. "Practicalities are not the issue," Quammen claims, when "flickering candles of life" are at stake; "saying no to the inevitable is one of the few precious ways our own species redeems itself from oblivion—or at least tries to" (1988, 112–13).

But let us back up a moment. To write that "mankind is only et cetera" in a sentence about planetary doom is to pass over a disturbing thought quickly; however, Quammen does return to it, in a passage conveying not only his strongest statement on human mistreatment of the Other but also his genuine love of humanity. His subject is the Gaia hypothesis, lately winning the enthusiasm of many New Agers but not many scientists, despite the respectable reputations of some of its promoters. Developed and popularized by J. E. Lovelock and others, this theory holds that Gaia—the ancient Greek version of Mother Earth—is so wonderfully made as to be able to heal herself of any and all climatic, geological, biological, botanical (and so forth) disasters, no matter what their source. Having mentioned that one of Gaia's ingredients for keeping the atmosphere balanced is methane gas, which results from decaying marsh plants and food digesting in the bodies of large animals, Quammen does not mince words as he sums up his opinion of this latest brand of scientific optimism:

> Evidently this Gaia . . is as cold and as Olympian a bitch as any goddess that man ever dreamed into being. She will endure as she has endured, she will provide as she has provided, she will cure herself of whatever damage humanity may inflict. Which is good, I suppose. It's grounds for a certain stoic and hyperopic sort of satisfaction. Okay, well, whoopee. Life on earth will continue . . . whether humankind cooperates, or the contrary.
>
> But the question that nags me is this. When humanity's earthly misbehavior has progressed to the point where even our farts can't redeem us, won't Gaia simply cure herself of *Homo sapiens*? (1988, 160)

Shock tactics, yes, but this criticism of "humanity's earthly misbehavior" resulting from homocentrism is properly understood only in the context of Quammen's larger agenda: to encourage perception of ourselves as part, and not necessarily in control,[3] of the entire natural scheme of things, as a product of the same evolutionary processes that gave rise to all other species. "How should a human behave toward the members of other living species?" becomes an examination, first, of how he or she should *not* behave. A certain amount of blasting away of old foundations is necessary if new construction is to get under way, if a fuller and more satisfying experience of nature by humans—and of humans by nature—is to be achieved.

All of nature. Not just the "white-tail deer or the parakeet or the puppy" mentioned above—that is, not just the animals for whom we feel easy affection. This brings us to Quammen's problem with the animal rights camp, with which one might initially expect him to side. "Animal Rights and Beyond" discusses a fairly old debate given new force in the 1970s and 1980s by Peter Singer and Tom Regan.[4] Although Quammen admires their attempts to grapple with a philosophical and practical issue that others dismiss too readily, he faults them for "smugness and myopia not too dissimilar to the brand they so forcefully condemn" (1985, 140). Because Singer, Regan, and company have to leave human beings the option of eating *something* while refraining from violating the creaturehood of other things, they have had to specify which creatures deserve to be left alone, and which do not. With sentience and the capacity to suffer (presumably in the recognizable, noisy fashion of humans, who cry or cry out) as the chief criteria, the vegetable kingdom loses the game entirely.[5] Plants have no brain, no central nervous system, no vocal chords. Quammen thus finds animal rights ethics too limiting—too exclusively animal, if what these thinkers are truly after is a "new moral framework" (from the essay's subtitle). Despite these advocates' attacks on speciesism and human chauvinism, their own moral equations are for Quammen "decidedly anthropocentric. Make no mistake: Man is still the measure, for Singer and Regan. The test for inherent value has changed only slightly [since Protagoras]. Instead of asking *Is the creature a human?*, they simply ask *How similar to human is similar enough?*" To which Quammen responds, "some simple minds would say: Life is life" (1985, 140). Existence itself confers value; life—and not its human form alone—is the final arbiter of significance.

Well, how *should* a human behave toward the members of other living species?

Not all human impact on nature and natural processes comes under bitter fire in Quammen's work; he is much more forgiving when a man or woman makes some attempt to look in the eye—not just metaphorically but literally—the creature he or she finds it necessary to destroy, whether for a meal or for some other purpose. Quammen makes this point early in *Flight* with a story about his own horrific discovery one evening of hordes of black widow hatchlings in his study; having dispatched them with a can of Raid, he has had them on his mind ever since: "The face of a spider is unlike anything else a human will ever see. The word 'ugly' doesn't even begin to serve. 'Grotesque' and 'menacing' are too mild." Unable to give good "reasons" why a spider has (too) many eyes and (too) many legs, Quammen adds:

> I only know that, when I make eye contact with one, I feel a deep physical shudder of revulsion, and of fear, and of fascination; and I am reminded that the human style of face is only one accidental pattern among many, some of the others being quite drastically different. I remember that we aren't alone. I remember that we are the norm of goodness and comeliness only to ourselves. I wonder about how ugly I look to the spider. (1988, 7–8)

What he did not do that evening but now forces himself to do—and urges readers to do, in imperative verbs tempered with incongruous images—is to "make eye contact with the beast, the Other, before [deciding] upon action. No kidding, now, I mean get down on your hands and knees right there in the vegetable garden, and look that snail in the face. Lock eyes with that bull snake. Trade stares with the carp. . . . *Then* kill if you will, or if it seems you must." Despite the lighthearted language early in this passage, it shares with much post-Thoreauvian nature writing a serious commitment to the ideal of personal encounter with nature's entities; one thinks of the mysticism of a Gary Snyder or an Annie Dillard. But true to his humbler persona, Quammen offers the disclaimer that looking the Other in the eye has not necessarily "made [him] gentle or holy or put [him] in tune with the cosmic hum, but definitely it has been interesting" (1988, 7).

With this story Quammen begins to do two things in *Flight of the Iguana* that he did to a much lesser extent in *Natural Acts*: to imagine another creature's perspective as a corrective to the human one, and to involve himself—his past, his foibles, his idiosyncrasies—in the issues at hand. Thus we increasingly find Quammen the man as a character in his own natural history, a persona willing to temper his potentially dogmatic statements about right relations between humans and nature with admissions of his own culpability. To live in the Western culture of the twentieth century is necessarily to be implicated, author and character seem to suggest, in a vast array of domineering, anthropocentric activity; but it may be possible to relinquish—consciously and conscientiously—some of that dominance, one instance at a time.

Quammen remarks several such instances in the lives of others besides himself, such as the crew members of a Soviet ice-breaking ship who went out of their way one winter to rescue "a thousand desperate whales" trapped by ice in the Senyavina Strait. The episode is especially noteworthy given the Soviets' frequent violation of quotas set by the International Whaling Commission. In this case they chose to break a path through the ice and draw the dying belugas out to the open sea by means of classical music played over a loudspeaker—when they might more easily have made a lucrative killing. Quammen wonders whether "there had come a sudden new Soviet recognition of kinship with the cetaceans," but he tells us he was never able to find out for sure (1988, 123).

This kinship ideal is the heart of Quammen's brand of natural history. Of all his statements regarding kinship between humans and nonhumans —based on ontological arguments and on ecophilosophical ethics—the strongest appears in *Flight*'s "Talk Is Cheap: A Personal Message from Washoe the Chimp."

With the help of a book by Eugene Linden (1986) on language experiments conducted with chimpanzees, Quammen here entertains "new and careful thoughts on . . . the nature of personhood." After relating the story of a chimp (Washoe) who learned to communicate "cognitively" by means of American Sign Language with her human guardians, including graduate student Roger Fouts, Quammen goes on to blur the line between humans and nonhumans more radically than most have been willing to do (1988, 114–15). "*What is a person?*" he asks, following up with a star-studded list of Western thinkers who have also posed the question, right up to Darwin (with his concept of "continuity and incremental transformation") and contemporary sociobiologist E. O. Wilson. Quammen's own answer:

> "Person" is just a word, after all . . . [but] an eloquent word, a richly connotative word, and one well suited for use in exactly that foggy no-man's-land between humanity and the rest of the biological community. I think it's a word that is wasted if judged to be merely a synonym for *Homo sapiens*. To me it seems that a person is any creature with whom you—or I, or Roger Fouts—can have a heartfelt and mutual relationship.(1988, 20)[6]

To instill a sense of kinship in readers toward the subjects of one's natural history by remarking similar "advanced" capabilities (cognitive use of language), similar tastes (for classical music), or similar physiology (the human eye and that of the octopus) seems logical strategy. Just as logical but more difficult to manage with virtuosity is the anthropomorphic metaphor: writing of the nonhuman in terms of the human has long been a way for nature writers to help readers identify with various plants and animals, especially the more bizarre ones—those more apparently removed from human experience. Such a strategy poorly handled, however, can misconstrue and sentimentalize natural facts, as John Burroughs demonstrated in his famous 1903 attack on "nature fakers."[7] Used more subtly, anthropomorphic metaphors can help to galvanize overt statements concerning kinship, ultimately serving the antianthropocentric ethic I have sketched as ecocentrism.

In this connection we can now look more closely at Quammen's essay on the sea cucumber, that "homely gob of living matter," which, we may be surprised to find, "ha[s] one on us," the self-styled higher animals (1985, 7). Although these ungainly, unbeautiful creatures "have never heard of eyes"—being unaware, we are to understand, of "any consensus . . . as to how an animal [is] supposed to behave" (1985, 4) they have nevertheless evolved the curious and highly useful skill of evisceration, of turning themselves inside out and expelling their own innards whenever the insidious parasite called a pearlfish enters the anus and "coolly set[s] up housekeeping" (1985, 6). Like the sea cucumber, male human beings "so foolhardy as to urinate while bathing naked" (1985, 7) in Amazonian waters are subject to another tiny swimmer, the candiru, which is to a man what the pearlfish is to the sea cucumber—a fishy parasite "shameless in the liberties taken, the indignities inflicted" (1985, 6). If you do venture such a skinny-dip, Quammen advises, "during submersion, hold the end" (1985, 7). Whereas a sea cucumber can, like its cousin the starfish, regenerate organs lost when it violently flushes the pearlfish out of itself, in men the only cure for invasion by the candiru is amputation—"In which event," Quammen delicately points out, "there is no question of regeneration" (1985, 7).

One of dozens of anthropomorphic sleights-of-hand in Quammen's work, this extended comparison between the sea cucumber and the human animal is old-style natural history insofar as it teaches us about a marine organism few of us will ever see in life. But it is new-style natural history in two ways: not only does Quammen spend most of his pages making the organism "familiar" by describing a medical condition it sometimes shares with humans, he also deliberately lets the so-called lower animal upstage the so-called higher one.

In every appearance of anthropomorphic metaphors in both *Natural Acts* and *Flight*, familiarity is the chief goal, with humor running a close second and humility (as in the sea cucumber essay) on the agenda often enough to carve a significant place for itself in readers' minds. Some are so simple as the quick simile: "*Miastor* [gall midge] daughters cannibalize the mother from the inside, with ruthless impatience, until her hollowed-out skin splits open like the door of an overcrowded nursery" (1985, 170). In another essay the second-person pronoun aids Quammen's metaphor in driving home the fact of "hermaphroditic but not self-fertilizing" earthworms: "imagine having a full sister whose mother was your father" (1988, 16). Personification of one kind or another spices the majority of the essays in each book: bedbugs are "disreputable," "sly," "sneaky," "libertine" (1988, 28–30); scorpions are "shy" (1988, 43); Venus flytraps are "choos[y]" eaters (1988, 49); crows are "bored" with their station in life, and therefore given to "detachment and cruel humor," even "drug abuse" (that is, they enjoy the formic acid on their feathers resulting from a roll in an ant colony [1985, 30–35]). Finally, Quammen sometimes gives us a reverse-anthropomorphic or zoomorphic metaphor, as when he compliments "the romantic imagination of mankind" by calling it "a hidden animal, a wondrous and inextinguishable beast" (1988, 96). In the end, the special achievement of these books purporting to be about close encounters of the strangest kinds is that before we quite know how it has happened, everything has come to look pretty recognizable, after all.

As a master of the anthropomorphic touch used to encourage a sense of connectedness between reading subject and nonhuman object, Quammen falls in with good company. J. Baird Callicott points out, for example, that such metaphors in Aldo Leopold's *Sand County Almanac* (1949), while "always restrained by and confined to the ethological facts of the animal behavior he describes," are nevertheless able "to excite our sympathy and fellow-feeling by portraying animal behavior as in many ways similar to our own and as motivated by similar psychological experiences" (1989, 125). Rachel Carson, Leopold's contemporary, was among those few nature writers who could take similar motivations so far as to make "characters" of her objects of study, delving into a sanderling's or a mackerel's mind (or what passes for its mind) without sacrificing scientific fact or reader credulity (see her *Under the Sea-Wind* [1955]). Like Leopold and Quammen, Carson believed some anthropomorphism was necessary to establish a feeling of relatedness between the creatures in her books and her human readers (Brooks 1972, 34).

Of course, the anthropomorphic metaphor works so well precisely because anthropocentrism itself comes to us quite "naturally." Here we come to a curious irony: the same century of ecocentric thought that challenges the human preoccupation with itself and its own ends has produced a body of popular natural history in which the human observer/participant looms larger than ever before. As in other current intellectual arenas, recent nature writing may obliquely present the author or persona who is "doing" something—in this case, studying nature—as himself a condition of the thing he is approaching, of the reality constituted not only by the object of study but also by the act of examining it.[8]

In effect, this represents an enlargement of Norwegian philosopher Arne Naess's description of late twentieth-century "deep ecology," which rejects any "man-in-environment image in favour of the *relational, total-field image*." Deep ecology insists that "organisms are knots in the biospherical net or field of intrinsic relations. An intrinsic relation between two things *A* and *B* such that the relation belongs to the definitions or basic constitutions of *A* and *B*, so that without relation, *A* and *B* are no longer the same things" (1973, 95). These organisms include, of course, *Homo sapiens* as well as all other living things. So if we let *A* equal the observer and *B* equal the observed, then the omnipresence of Quammen-as-character in his nature essays, particularly in *Flight*, becomes ecologically significant as well as humanly interesting.

The personal narratives may be brief—as in that picture of Quammen discovering his office full of black widow spiders, or a reference to himself as a "large pink creature . . . with sunburned ears" watching an iguana swim ashore (1988, 166), and that inclusion of himself (more than a little nervous) among a party of Central Americans crossing the border illegally from Mexico into the United States. Or personal narrative may share center stage with natural history, especially toward the last half of his second book. Georgia's Okefenokee Swamp becomes a metaphor (as well as the locale) for an important reunion between Quammen and a buddy with whom he shared the now-lost spirit of the 1960s (1988, 241–60). A stenothermal creek in Montana symbolizes for him another, irrevocable loss, this time friends who divorced each other (1988, 270–77). And the chambered nautilus—which gradually accumulates layers of shell to form spaces for storing seawater or gas to help the creature sink or rise—provides Quammen with a poignant emblem of the mysteries of time and place, particularly his youthful days spent in the hometown of William Faulkner, consummate novelist of time and place (1988, 261–69).

With such vignettes lacing Quammen's larger discussion of nature and the human link, any apparent reductivism in the distinctions he makes—or, more often, dispels—between humans and nonhumans, self and Other, subject and object, becomes emphatically that—apparent only. But if his blend of these heretofore sharp dichotomies is not simplistic, neither is it haphazard or sloppy. Rather, each holds the other in special poise: again, each contributes to the definition of the other, as in Naess's relational equation—or, more poetically, as in the figure of yin and yang Quammen sees etched into the landscape of the Tularosa Basin (1985, 193–208).

An anthropocentric stance toward nature is not a simple question of human ignorance, nor is it always (or even often) a matter of malicious thinking and behavior—no more than ecocentrism is another word for misanthropy; it is not. Rather, the continuum between these concepts is like so many others involving personal and cultural values. Given a chance and a reason to see a person, a place, or a sea cucumber differently, I may come to value him or her or it as I did not before, chiefly because I know and understand as I did not before. As my view of the universe widens, so deepens my "sense of wonder," to recall the title of Rachel Carson's last book (1965). To educate us both scientifically and ethically in the wider view, to develop in us a greater sense of wonder than we could ever achieve looking only at ourselves and each other: this is what the American descendants of Thoreau, such as David Quammen, have been quietly up to. Whereas Emerson warned that natural history alone is "barren," but if "marr[ied] to human history" becomes "full of life" ([1836] 1983, 21), perhaps we are now in a position to consider the reverse, or something like it: if not actually barren, human history divorced from natural history may well be grossly incomplete.

NOTES

1. An especially interesting example is that of Joseph Wood Krutch, drama critic turned nature philosopher. Thomas Lyon notes that Krutch's later work represents a 180-degree movement from the position taken in his early Modern Temper (1929), that humankind and nature are "fundamentally antithetical" (1989, 79).

2. David Ehrenfeld writes, "It is a serious mistake to assume that because we are at present the most conspicuous creation of Nature, each of her other myriad creatures and workings can somehow be turned to our benefit if we find the key. As conservationists use it, this is one of the most gentle and well-meaning of the humanistic deceptions, but falsehoods that spring from good intentions are still falsehoods" (1981, 199).

3. For an entertaining discussion of the two meanings behind the phrase "the control of nature," see McPhee (1989).

4. Australian Peter Singer and American Tom Regan are the authors, respectively, of *Animal Liberation* (1975) and *The Case for Animal Rights* (1983), among other works.

5. For more on this, particularly the animal liberationists' misconstruction of pain and their poor equation of wild and domestic animals, see Callicott (1989).

6. Similarly, Joseph Wood Krutch points out the developed consciousness—a kind of personhood—an animal shows when it "has been really accepted as a companion. Not only cats and dogs but much less likely animals seem to undergo a transformation analogous to that of human beings who are introduced to a more intellectual, more cultivated, more polished society than that in which they grew up" (1978, 123–24).

7. Burroughs takes to task his contemporary nature writers for anthropomorphizing their material so far as to falsify observed and observable facts.

8. As John D. Barrow and Frank J. Tipler (among many others) point out, this is not an exclusively twentieth-century phenomenon; at least since Copernicus Western science (and more recently, the humanities) has understood that "our picture of the Universe and its laws are [*sic*] influenced by an unavoidable selection effect—that of our own existence" (1986, xi). But this idea has come to the fore in recent decades with the advent of quantum mechanics and other scientific developments, discussed throughout Barrow and Tipler's book.

REFERENCES

Ackerman, D. 1990. A Reporter at Large: Whales. *New Yorker*, 26 February.

Barrow, J. D., and F. J. Tipler. 1986. *The Anthropic Cosmological Principle*. Oxford: Clarendon Press, New York: Oxford Univ. Press.

Brooks, P. 1972. *The House of Life: Rachel Carson at Work, with Selections from Her Writings*. Boston: Houghton Mifflin.

Burroughs, J. 1903. Real and Sham Natural History. *Atlantic Monthly*, March.

Callicott, J. B. 1989. *In Defense of the Land Ethic: Essays in Environmental Philosophy*. Albany: State Univ. of New York.

Carson, R. 1955. *Under the Sea-Wind: A Naturalist's Picture of Ocean Life*. New York: New American Library.

———. 1965. *The Sense of Wonder*. New York: Harper and Row.

Ehrenfeld, D. 1981. *The Arrogance of Humanism*. New York: Oxford Univ. Press.

Emerson, R. W. [1836] 1983. Nature. In *Ralph Waldo Emerson: Essays and Lectures*. New York: Library of America.

Gould, S. J. 1990. The Golden Rule: A Proper Scale for Our Environmental Crisis. *Natural History*, September.

Krutch, J. W. 1929. *The Modern Temper: A Study and a Confession*. New York: Harcourt, Brace.

————. 1978. *The Great Chain of Life*. Boston: Houghton Mifflin.

Leopold, A. 1949. *A Sand County Almanac, and Sketches Here and There*. New York: Oxford Univ. Press.

Linden, E. 1986. *Silent Partners: The Legacy of the Ape Language Experiments*. New York: Times Books.

Lovelock, J. E. 1979. *Gaia: A New Look at Life on Earth*. Oxford: Oxford Univ. Press.

Lyon, T. 1989. *This Incomperable Lande: A Book of American Nature Writing*. Boston: Houghton Mifflin.

McPhee, J. 1971. *Encounters with the Archdruid*. New York: Farrar, Straus, and Giroux.

————. 1989. *The Control of Nature*. New York: Farrar, Straus, and Giroux.

Naess, A. 1973. The Shallow and the Deep, Long-Range Ecology Movement: A Summary. *Inquiry* 16: 95–100.

Quammen, D. 1985. *Natural Acts: A Sidelong View of Science and Nature*. New York: Nick Lyons Books.

————. 1988. *Flight of the Iguana: A Sidelong View of Science and Nature*. New York: Delacorte Press.

Regan, T. 1983. *The Case for Animal Rights*. Berkeley: Univ. of California Press.

Singer, P. 1975. *Animal Liberation: A New Ethic for Our Treatment of Animals*. New York: A New York Review Book.

Turner, F., Panelist. 1990. Only Man's Presence Can Save Nature. *Harper's*, April.

Omni Meets Feynman: The Interaction

Between Popular and Scientific Cultures

DAVID A. STONE

When scientists think of the public, especially when they express their frustration at the difficulties of communicating with a general audience, they tend to characterize their readers in terms of absences: the low level of public education in science; the public's intolerance for the technicalities of scientific discourse; the public's inability to accept that, regarding many of the burning questions of the day, scientific knowledge is incomplete and gives ambiguous answers. What is easily overlooked is that the science-reading portion of the public has its own culture, with a fairly coherent system of positive beliefs and values concerning science as well as the negative ones just mentioned.

To elucidate how scientific and popular cultures clash in the realm of science, I examine passages from two authors, one from each culture. *The Feynman Lectures on Physics* (1963), by the late Richard Feynman, represents scientific culture, and on the side of popular culture I use an article from *Omni* entitled "Into the Woods" (1988), by Jessica Maxwell. Analysis of these two works indicates that the Scientific Revolution is not yet over. The relative power of the two parties, scientific and antiscientific, has shifted, and the battleground and weapons have changed, but something like a guerrilla war is still being fought. The scientific community plays the role of police and army, while the members of the popular, antiscientific culture are the guerrillas, retreating when directly attacked, but pouncing on their enemy's every weakness.[1]

Thus scientists who wish to communicate to the general public must do so over a wall of hostility and amid a barrage of propaganda and counterpropaganda. This is what makes their task so difficult. It is clearly necessary for the good of science and for the public welfare to bring about a rapprochement between the scientific and popular cultures. Scientists will have to find a way to listen sympathetically to the point of view of popular culture. In the final portion of this essay I shall discuss what I believe must be the first step for scientists to take in this direction, as well as examples of scientists who have succeeded in addressing the popular culture.

THE POPULAR CULTURE OF *OMNI*

Jessica Maxwell's "Into the Woods" describes her search for exotic animals—some so exotic that, like the beasts in a medieval bestiary, they may not even exist—and her reactions to them when she finds them. Her final paragraph provides a fair idea of the content and style of the whole article:

> If you ever find yourself discontented about our age, thinking that technology's homogenization has bankrupted the planet of its original eccentricities; that the presence of McDonald's in Beijing and of Whitney Houston T-shirts in Borneo has rendered the earth, once again, quite flat; then it is helpful to recall the day of the stiff dogs, to remember that each spring ice worms begin their tentative ascent through the thinning Alaskan snowpack, and brine shrimp bloom in the Great Salt Lake like a colossal aquatic Matisse. The new greenery sends inaudible shrieks of joy through the colonies of Texas leaf-cutter ants, and gold beetles huddle in wonder at the sudden profusion of morning glories. From Mexico to California, monarch butterflies plot their return trips north. The giant Pacific octopus slides along the bottom of Puget Sound, and gentle manatees float in Florida's waterways. For just beyond the roar of Mercs and Evinrudes, just beneath the dispassionate skins of Congoleum and Astro Turf, all strange symbols of modern life, that old natural magic is still going strong—and you don't have to float down the Amazon to find it. It's right here, snorting, flapping, wallowing, and burrowing in America's own backyard. (1988, 92)

Maxwell's article demonstrates that popular culture is not a culture of the uneducated or the unintelligent.[2] The difference between scientific and popular cultures is more subtle than that between knowledge and its absence. Popular culture is built not on ignorance but on different conceptions of which observations deserve the name of knowledge and of how to organize that knowledge. In *scientific* discourse, these ways of organizing knowledge are termed "subjective" and "objective," respectively, with the connotation by now inevitable that the latter is preferable. Since I do not want to take sides, I avoid these terms and refer to the *knowledge structures* of scientific and popular cultures. Scientific knowledge is structured by logic, cross-references, and consensus. Popular knowledge is structured by each person in terms of his or her own psychological responses to it: the emotional and imaginative states experienced at the moment of reading a story, for example, and the relationships developed with the author and the elements of the story being told.

From the scientific point of view, Maxwell's knowledge is desultory: "When the mountains came barreling out of the seas, they [the manatee species] were here. When glaciers took the land like a cold cancer, they were here. What are the pyramids held against the cellular memories of manatees?" (91). The speeds at which mountains rise out of the seas and glaciers spread across the land are the reverse, in the context of this chronology, of her analogies: glaciers come and go in thousands of years, while mountain ranges take millions to form and erode. It is the lack of logic-based, systematic interconnections in knowledge that makes Maxwell's knowledge scientifically unintellectual, not the lack of knowledge itself.

But the readers of *Omni*, though they may be unintellectual, are not timid. Like the medieval readers of a bestiary, they are eager to travel quite far in imagination, with a daring that may leave readers of Feynman's physics feeling queasy. Like a bestiary, which contains information from natural history, religious mythology, folktales, and even etymological conjectures (White 1954, 253–59), so Maxwell's article quotes (in full) a poem about ice worms by Robert Service and repeats a folktale about baker ants:

> There are even ants in Texas that plant rice, harvest it, crack it into meal, then make cakes out of it and set them in the sun to bake. Or are there? . . . "Good stories die hard," [Sanford] Porter [a research associate in zoology at the University of Texas] says with a chuckle. "Those harvester ants were first reported back about 1860." The story got into an encyclopedia and spread around the world. It was disproved early this century and has been disproved several times since then. (44)

By the standards of modern science, then, *Omni*, like a bestiary, seems to mix a confused jumble of facts with transparently false fables. But it is equally the case that the modern scientific organization of knowledge according to scientific methodologies would be a confused jumble to the twelfth-century readers of a bestiary, or to their present-day successors, the members of popular culture. Which of these two assessments is correct? I believe that each is more or less justifiable—but only by reference to its own belief system. There is no "natural" way to codify knowledge. We learn to do so within our native culture, whatever it may be; once we have been thoroughly trained, we can conceive of no other way, and we call that way "natural." It *is* natural—to oneself and one's compeers, but not necessarily so to others.

Scientists are apt to regard popular culture (in its attitude toward science) as abnormal and ignorant. This is a prejudice they will have to overcome if they wish to address a popular audience. They will also have to cope with the reverse prejudice. From the viewpoint of popular culture, it is the scientific community that is abnormal and ignores crucial questions about how the scientific enterprise should be conducted and its discoveries put to use.

THE SCHIZOPHRENIC SPLIT IN SCIENTIFIC CULTURE

The Scientific Revolution took place in particular historical circumstances. It was neither natural nor inevitable for men to think of combining two antithetical modes of inquiry. One mode, held also by the Greek philosophers, the bestiarists, and by modern popular culture, was motivated by the eager curiosity and thirst to make sense of the universe as a whole in terms of what was already familiar. A typical and very important problem in this mode of thought is to define humankind's place in the universe in some way that justifies our natural pride in being human.

It may be objected that this problem is, in the first place, not a scientific one, and, second, is not particularly important to ordinary men and women, such as the readers of *Omni*. The first objection begs the question: it represents the position of scientific culture on this issue while ignoring the views of popular culture. As to the second objection, it is true that in ordinary times, ordinary people do not trouble themselves about such large-scale questions as their position in the universe; but this is because whatever belief they inarticulately and half-consciously hold is not being called into question. The strengths of their beliefs only become apparent when they are challenged. The Copernican, Darwinian, and Freudian shifts of paradigm, for example, aroused violent opposition because they took away from Western man something on which he prided himself: his location at the center of the universe in the first case, his status as God's specially created creature in the second, and (most recently) his self-definition as the uniquely rational being. Nowadays we justify our feelings of importance as a species by referring to our supremacy in intelligence and creativity; but let computers approach equality with us in these respects—as is already starting to happen—and another very profound social crisis will occur.[3]

The second mode of inquiry of the Scientific Revolution derived from the plodding methods of artisans—watchmakers, shipwrights, and lens-grinders—whose techniques were less "natural" than the phenomena to which they were applied by science. The two methodologies of modern science, inspiration and perspiration, have still not been truly united,[4] as can be seen by analyzing the following passage by Feynman from the point of view of popular culture:

> The things with which we concern ourselves in science appear in myriad forms, and with a multitude of attributes. For example, if we stand on the shore and look at the sea, we see the water, the waves breaking, the foam, the sloshing motion of the water, the sound, the air, the winds and the clouds, the sun and the blue sky, and light; there is sand and there are rocks of various hardness and permanence, color and texture. There are animals and seaweed, hunger and disease, and the observer on the beach; there may be even happiness and thought. Any other spot in nature has a similar variety of things and influences. It is always as complicated as that, no matter where it is. Curiosity demands that we ask questions, that we try to put things together and try to understand this multitude of aspects as perhaps resulting from the action of a relatively small number of elemental things and forces acting in an infinite variety of combinations. (1963, 2-1)[5]

Jessica Maxwell (to judge by my first quotation from her article)—indeed, any member of popular culture—would agree with Feynman as far as "curiosity demands that we ask questions." Then, "that we try to put things together," might mean no more than is expressed in the common phrase "putting two and two together to make four"; and who could disapprove of that kind of mental exercise? Only with the third version of the challenge, "to understand this multitude of aspects," does Feynman express the values of scientific culture. And it is by no means clear that members of popular culture would agree with Feynman's last statement, for this contains the germ of the contemporary scientific definition of man as a piece of biochemical machinery; and what pride can we take in that? In fact, with his last clause—"the action of a relatively small number of elemental things and forces acting in an infinite variety of combinations"— Feynman has deftly leaped from the general, essentially popular, mode of thinking about nature to the specifically scientific one. As members of the popular culture, we may have been distracted by Feynman's rhetorical skill, even perhaps by our own familiarity with the discourse of scientific culture, from noticing how wide a chasm he has just bridged.

Once we start to look for what has been omitted from the Feynman passage, we find more to disturb us. What are we to make of the discrepancy between Feynman's description of himself as an observer on the beach, compelled by "curiosity" to ask questions about Nature, and his actual but unmentioned status as an important member of the community of physicists? (In the days when such a community hardly existed, Newton might legitimately have described himself in terms such as these—though Galileo would never have done so, knowing the importance of scientific investigation to society.) These remarks, to which many scientists would subscribe, were made by a Nobel Prize winner, a man who spent most of his career in the center of the community of physicists, a man who made his name as a member of the Manhattan Project team, which made the first atomic bomb. It takes a profound schizophrenia for a scientist to be able to describe what it is that he does in terms as sincere and convincing as Feynman's. I shall argue later that the training required of aspirants to membership in the scientific community tends to induce in them the same cultural schizophrenia.

Thus, from the standpoint of popular culture, what is abnormal—what needs to be accounted for—is the cultural newcomer, the recently invented scientific view. Invented for good reason, no doubt; profoundly useful— and profoundly affecting what we mean by the term *useful*[6]—but to this day, not a viewpoint that is easily absorbed even by the greatest scientists, far less by the man in the street.

THE OBJECT OF KNOWLEDGE IN POPULAR
AND SCIENTIFIC CULTURES

When it comes to the description of nature, a culture's deepest-held values are embodied in what it perceives to be the ultimate "realities."

These entities form the building blocks of the culture's way of explaining the universe, the things to which other things are likened or from which they are distinguished. Whatever cannot be compared to the fundamental "realities" is either ignored or categorized as "meaningless," with the connotation of "illusory" or "insignificant." Both scientific culture and popular culture (as it regards science) are concerned with nature and humans' relationship to nature. But the two cultures have rather different ideas about the "reality" of these objects of their inquiries.

In the world of *Omni*, to recall Maxwell, Nature's productions are more interesting, less "flat" than those of humankind: "If you ever find yourself discontented about our age, thinking that technology's homogenization has bankrupted the planet of its original eccentricities; that the presence of McDonald's in Beijing and of Whitney Houston T-shirts in Borneo has rendered the earth, once again, quite flat," just think of ice worms or brine shrimp. Thus, one purpose of Maxwell's article is to unflatten life a bit and thereby to attempt to join Nature in opposition to "technology's homogenization." But the attempt to join in Nature's playfulness cannot be fully realized because we are not wholly of (*Omni*'s) Nature. Look at how idyllically the manatees' social existence is described, like that of infancy or Eden:

> They have no natural enemies besides man, no social hierarchy, no daily set routines. Many of them will swim right up to divers to be petted. When manatees are together in groups, one manatee does not dominate the herd, although a manatee often initiates playtime. Manatees love to play. Between somersaults, headstands, tail stands, and barrel rolls, they bodysurf, kiss, nuzzle, bump, and chase one another. And eat. (90)

But the author and her readers stand outside this secular paradise. Adult humans can at best observe nature nostalgically: "With a sad smile, she adds, 'They really ask for nothing. They don't kill anything. They just want to float along—and eat'" (92).

If we cannot join in this idyllic Nature, what place in the universe is left for us, according to Maxwell? For one thing, we stand far enough outside Nature to be her audience, to respond with enthusiasm and imagination to her marvelous tricks. But we have also another part to play in the universe, according to medieval and popular cultures. We separated ourselves from Nature; we gave up the innocence of elephants and manatees. In exchange we acquired the knowledge of good and evil—especially scientific knowledge.

"Goodness," in both the culture of the medieval bestiary and the popular culture of *Omni*, is correlated with membership in the community. "Good" scientists, those *Omni* likes, are introduced to Maxwell's readers by name, position, and sometimes by physical description. They address us, through her, in their own words, and thus establish their connection to the general public. The following quotation demonstrates, on the other hand, that when power emanates from a source with which Maxwell feels no.psychological connection, she is deeply hostile to it and regards it as "evil": "Would computer wizard Steve Wozniak keep his teeth in his stomach, and his stomach in his head? or replace a lost eye with a nose? The Maine lobster does. If a large gator attacked microchip genie Bill Gates, could Gates cast off his leg, then grow another? The starfish can. This alone is reason enough for technobrats to remember to take a field trip every once in a while" (43). Popular culture allies itself with Nature, the performer of "white" magic; whereas the "technobrats" are dangerous because they are not members of the general community. The fact that these scientists have not only formed their own community but have even voluntarily rejected the companionship of the general public only makes them more dangerous in the eyes of popular culture.

Thus the primary realities of popular culture are those of power and connectedness. Power is "good" if it emanates from a source connected with the general community, and "bad" otherwise. The technicalities of how to manipulate the forces of nature and the factual dimension, from truth to falsehood, of scientific material are of little significance to popular culture.

In scientific culture, on the other hand, what matters most, what is "real," is whatever can be measured. More than that, it must be possible to foresee the possibility of controlling (within limits) that measurement, to produce whatever value is chosen. Science is humankind's quest for the ability to harness more and more natural phenomena to our chariot. From this point of view, all knowledge that may lead to this kind of control is valuable. The possibility, raised in the medieval bestiary and *Omni*, that humans might be in some ways distinguished from nature is never raised in scientific culture. Feynman's observer on the beach blends with the nature he observes: "There are animals and seaweed, hunger and disease, and the observer on the beach; there may be even happiness and thought."

The problem, in popular culture, of delineating the circumstances under which it is right for human beings to use their power over nature does not exist in scientific culture. Here is Feynman again: "[F]ar more marvelous is the truth than any artists of the past imagined! . . . What men are poets who can speak of Jupiter if he were like a man, but if he is an immense spinning sphere of methane and ammonia must be silent?" (1963, 3-6). To Feynman, the description of Jupiter in physicochemical terms is *the* truth because spinning spheres, methane, and ammonia can be controlled in the sense I just defined; whereas the ways in which Jupiter resembles a man are exactly those which *cannot* be expressed in mechanical or quantifiable terms. In popular culture a superhuman being who possesses power far beyond what is biologically plausible is a stock figure of science fiction, which concerns itself not with the details of how this power is generated and controlled but with problems arising from the good and harm done by such a being—"good" and "harm" according to the standards of brotherhood and decency of the common people, whether characters in the fiction or its readers. Does this not define Jupiter, the god? And does it not also fit the "mad scientist" of the cinema and popular fiction: a man originally motivated to do good for humankind (usually) but who has gone astray due to his isolation from the community and the effect upon himself of his extraordinary power, until he becomes a menace to society (Basalla 1976)?

THE DISCOURSE OF SCIENTIFIC CULTURE

The power of scientific culture is bound up with its claim that science alone has access to a special kind of truth. What is special about scientific truth, supposedly, is its freedom from rhetorical tropes and from metaphysical and epistemological hypotheses. But this assertion is itself a rhetorical device;[7] in fact, the canons of scientific writing are designed to support these philosophical assertions, as we can see in the following quotation from Feynman:

What do we mean by "understanding" something? We can imagine that this complicated array of moving things which constitutes "the world" is something like a great chess game being played by the gods, and we are observers of the game. We do not know what the results of the game are; all we are allowed to do is to *watch* the playing. Of course, if we watch long enough, we may eventually catch on to a few of the rules. *The rules of the game* are what we mean by *fundamental physics*. (1963, 2-1)

In Feynman's metaphysics, as illustrated by this passage, the provenance of the laws of physics cannot be subjected to inquiry. They preexist humanity, since they belong to the realm of the gods. This point of view is very useful to science in its quest for authority. But it is not the whole truth. Feynman fails to incorporate a crucial aspect of how scientists work. The laws of physics, as best they are known at any time, are not decreed by gods; they are agreed upon by the community of physicists. It is from the consensus of physicists that these laws are refined and cleared of error; it is the community of physicists who are the "gods" playing Feynman's chess game. And Feynman is not only a mortal watching the gods at play—which is, no doubt, how he felt when he was privately engaged in struggling to understand nature—but also one of the gods himself.

The quotation from Feynman is also revealing of the epistemology of modern science. He combines two modes of discovering the laws of science. One is by means of logic: if we assume that the laws of physics are the rules of a game of chess, then there must be some logic to their structure, which we can try to analyze. The other is historical: if the game is played by gods, then it has been going on for a long, long time. Although our predecessor scientists may have made mistakes in their observations or analyses of the game, they were at least watching the same game that we now observe; and, to the extent that our analysis and theirs are accurate, they must therefore be consistent with each other.

Feynman's rhetoric is quite unmistakable and unlike that of any other scientist. On the whole, however, the rhetoric of textbooks and of scientific research has hardly changed since Newton's *Opticks*, except to become more impersonal and omniscient (Bazerman 1988). We are constantly subjected to an authorial voice that expresses neither much excitement nor much doubt, as befits a participant in the divine knowledge of the rules of the game.

The metaphysical, epistemological, and rhetorical postures I have just mentioned are pervasive in scientific writing, and have been so since the Scientific Revolution. In scientific research, proponents of opposing paradigms compete on the logical front by claiming greater theoretical rigor and better agreement with empirical evidence than their rivals can manage; they also compete on the historical front by claiming consistency with previous discoveries now thought to be secure (Latour 1987).

Textbooks reflect whatever is currently considered to be normal science. Part of their task is to demonstrate the success of the current paradigm in accounting for the natural phenomena of their chosen field. Another part is to teach a mythologized history, emphasizing whatever can be claimed as progress toward the current paradigm, ignoring or denigrating whatever cannot be assimilated to it, and generally proving that science continually makes systematic progress toward Truth. But this vision of science, which Thomas Kuhn (1970) calls normal science, is at best a caricature of true progress in science. All real progress in science comes about through paradigm shifts; that is, as a result of messy controversies which are as often superceded as resolved.

This mythologized view of the history of science is nonetheless absorbed by scientists-to-be, even though they soon learn in actual practice that science lurches from one controversy to another. Fortunately for their peace of mind, they learn to use the schizophrenic split, which I mentioned earlier, so that they can ignore the conflict between their experience of scientific history and the history they have learned (Latour 1987). When they do become aware of the conflict, most scientists probably argue that there is a difference between what happens to scientists, as individuals, and to science, as a branch of knowledge. Individual scientists are fallible, egotistical, pugnacious, greedy for power and prestige; they let themselves be unskeptically swept away by new ideas or cling to familiar ones despite their increasingly obvious inadequacy; but in the course of generations, these human lees are distilled away, and we are left with the pure truth of science.

While this analysis of the individual scientist contains some truth,[8] it retains the main function of the mythical history of science, which is to relegate scientific controversy to disputes between individual scientists, to a realm of discourse in which egalitarian values prevail, more or less, while making it impossible to find language in which to question the authority of science itself. Thus it is that scientific knowledge itself is reified and idealized, and science enforces its claim to be the sole mode of access to a Truth greater than that of any other epistemological system (Aronowitz 1988; Latour 1987; Mulkay 1979).

THE DISCOURSE OF SCIENCE IN POPULAR CULTURE

Popular culture brings its own metaphysics and epistemology to the interpretation of the discourse of scientific culture. In certain respects the fundamental beliefs of the two cultures are in accord. Where they differ, popular culture is marginalized, and its discourse about science can only be inferred from its comments about the discourse of scientific culture—and from its silences.

Like scientific culture, *Omni* on the whole expresses optimism about the future of humankind and faith that technology will make steady progress toward bringing that future to pass. Popular culture, again like scientific culture, protects its claim to authority from scrutiny by deflecting criticism onto individuals, so that it becomes impossible to challenge its social institutions. In its treatment of science, although there is little looking to the past, the future is mythologized in the same way that scientific culture mythologizes history. If problems caused by technology are raised, they are described as temporary anomalies or their import is muted by their being phrased in some hypothetical, science fictional context. This approach greatly weakens the possibility of pointing to the structural flaws in the technological basis of our actually existing society. Likewise, though individual scientists, as we have seen, are blamed for deviant behavior, there is no avenue for criticizing the scientific community as a whole. Thus popular culture acts in collusion with scientific culture to protect the social institutions of science.

It should not be surprising that popular and scientific cultures resemble each other in this aspect of their metaphysical systems. Both cultures are direct descendants of medieval culture;[9] however, while popular culture has remained relatively true to its feudal origins, scientific culture reworked its metaphysical foundations as it evolved during the rise of the bourgeoisie and the Enlightenment. According to the medieval system of metaphysics, the essence of the natural world, being divinely ordained, is perfect. Whatever is wrong is accidental, not essential—a transitory glitch. Thus, while a medieval bestiary might criticize human conduct, its suggestions for improvement are always addressed to sinners as individuals and are always expressed in moral terms (White 1954, 246). The church and the kingdom, the social institutions that represent the divine authority, are not to be touched.

The modern scientific point of view, which had to struggle for political and epistemological footholds in the time of Galileo and Sir Thomas Browne, has become so powerful that it controls language as well as providing society with the values and beliefs of its dominant sector. No belief system, be it communism or creationism, can hope for wide acceptance unless it justifies itself somehow by reference to science.[10] This means that the antiscientific component of popular culture cannot be articulated in a straightforward manner, and so it is easily disqualified as "nonscientific" by the dominant, pro-scientific culture. Thus popular culture cannot influence the discourse of scientific culture by any discourse of its own.

For example, given that Maxwell's article is about exotic animals, with which she seems to have had no previous empirical or theoretical experience, her reluctance to challenge the scientists she interviews is entirely appropriate. Only once does she mention doubting a scientist's word—and look how tentatively she does so:

> After ridiculing the idea that ants "bake rice bread," he [Roy Snelling, collections manager for the entomology section of the Natural History Museum of Los Angeles County] tells me about Southwest honey ants. "Just pop them into your mouth, holding on to the head and thorax, and eat them like grapes. They're delicious." . . .
>
> As convincing as Snelling was, I wasn't convinced. Not about the gastronomic joys of honey ants or the myth of Texas baker ants, . . . [h]aving read about baker ants in a book by a Texas naturalist who seemed to know his stuff. (44)

She protects herself in her challenge to Snelling not only by having another scientist's word to back her up (though he remains anonymous), but also by diluting her judgment of a matter of fact with an opinion about what is only a matter of taste—the delights of eating honey ants.

But there is stronger evidence of Maxwell's reluctance to adopt the discourse of scientific culture by judging facts and evaluating theories. She chose to write an article on several exotic animals rather than on a few creatures more familiar to her. Since the toads, octopuses, shrimp, and manatees are unrelated to each other (except through her interest), there can be no scientific theory for her to evaluate. And because these species were either previously unknown to her or were presented to her in novel circumstances, she participated in depriving herself of any authority to question matters of fact, within the discourse of scientific culture. That is to say, Maxwell has consented to the epistemology in which one accepts information from a higher authority, without the possibility of querying it.

Instead of entering the discourse of scientific culture, popular culture is therefore reduced to reacting to and commenting indirectly on it. Maxwell treats the scientific content of her article—which, in scientific discourse, is the essence of the matter—as no more than an opportunity to pursue her own agenda. Her "true or false" games and her lack of interest in the densely interlocked texture of scientific facts challenge not the truth of scientific discourse but its universality. She places her emphasis, as we have already seen, on her own feelings (compassion for the manatees and anger at "technobrats"), especially as they relate to issues of connection versus separation of power.

Maxwell can challenge the discourse of scientific culture only from *outside* the context it defines for itself. This is in large part because scientific culture is not open to hearing alternative discourse. The quotation from Feynman discussed above makes this clear: "What men are poets who can speak of Jupiter if he were like a man, but if he is an immense spinning sphere of methane and ammonia must be silent?" For in the silence following Feynman's rhetorical question, we hear the counterquestion: Why is scientific culture deaf to the truth contained in using Jupiter to symbolize humankind?

HEALING THE SPLIT BETWEEN SCIENTIFIC AND POPULAR CULTURES

I have expressed my belief that if scientific and popular cultures had a *realistic* understanding of their two histories, they would be better able to find common ground. How might such a rapprochement come about?

The first step would be to recognize that the mythologized history of science, with its distinction between science in the abstract and what it is that scientists do, is a fiction whose fundamental purpose is to establish a voice of absolute authority for science. The metaphysics of science supports this claim in two ways: first, by reifying science as something above the fallibility of individual scientists; and, second, by inculcating in scientists a schizophrenic split, whereby whatever doubts they may have are relegated to their private or unconscious lives, and thus can be categorized as inappropriate to scientific discussion. But as scientists come to recognize that there is no science independent of actual scientists, they will be able to question this ideology. Then they can recognize that the turbulent history of what scientists have actually done is all there is of science. They will have to acknowledge its blunders as well as its successes, its attempts to justify oppressive political ideologies as well as its potential for liberating humankind. They will realize that science's claim to be the sole repository of absolute truth is made by certain individuals (many of them scientists) who have an interest in ensuring that scientific authority remains dominant in our culture.

Once scientists are no longer dogmatically convinced of the absolute authority of science, they will be receptive to what popular culture has in common with scientific culture. In the previous sections I emphasized some of the conflicts between the two cultures; but there are also areas of agreement. The scientist's epistemology of empiricism is matched by the importance in popular culture of the evidence of the senses. For example, Maxwell takes care with her descriptions of animals and scenery, and she likes to report direct discourse, sometimes even including vocal timbres: "a fine whiskey baritone" (44) or "bell-like British tones" (46). To scientific culture's egalitarianism corresponds a democratic spirit in popular culture. Maxwell quotes not only professors and professional naturalists but also an Alaska Airlines clerk (44), an "old geezer" in a bar in Alaska (44), and a woman who does "just about everything" at Nature World, "a tourist attraction with the usual snack bar and tackola gift shop" (91).

The most important connection between popular and scientific cultures becomes clear once it is understood that popular culture is not intrinsically antiscientific; rather, it opposes the social and economic forces of rationalization. Recall again what Maxwell says of technology: "technology's homogenization has bankrupted the planet of its original eccentricities; . . . the presence of McDonald's in Beijing and of Whitney Houston T-shirts in Borneo has rendered the earth, once again, quite flat" (92). Popular culture mistrusts science when its particular methods of rational inquiry are harnessed by socioeconomic interests to serve their own purposes of rationalizing society. Science, especially normal science, and even more so the caricature of normal science that is presented as historical truth in the standard textbooks, does indeed have a component in its methodology that lends itself to this kind of abuse (Aronowitz 1988; Levins and Lewontin 1985). This is hardly surprising. Ever since Bacon (in the *New Atlantis*) prophesied the methodology of modern science leading and serving a rationalized society, science has grown in tandem with, and fostered by, the push of modern bourgeois society toward this form of dominance. The belief systems of both scientific culture and bourgeois capitalism combine mystical foundations with an emphasis on practical mastery; so it has always been easy for capitalism to subvert scientific values to its own ends (Handlin 1967; Weil 1967).

However, not all of science can be co-opted by the modern bureaucratic state. Revolutionary science, by overthrowing old paradigms and initiating the search for new ones, can be emancipatory, not only because it offers humanity new sources of power with which it might perhaps oppose the state, but also because it represents symbolically the ability of human creativity and unpredictability to throw off the harness of rationalization. For instance, much has been written by nonscientists about the moral implications of quantum mechanics and the theory of relativity for human free will. In the discourse of scientific culture, most of this writing is an unwarranted generalization of fundamental physics, and is therefore meaningless. But in the view of popular culture it is crucial to claim the support of science, however metaphorically, because, as I have said, popular culture has no other way to make heard its protest against rationalized state power.

Among scientists who write for the general public there may be a few who have no further goal than to preach the doctrines of scientific culture in hopes of encouraging the faithful and converting some doubters. But I suppose that the majority are in a more ambiguous position and wish to open a genuine dialogue between scientific and popular cultures. That they have such a wish indicates that they feel, however unconsciously, some affinity for popular culture and its mistrust of scientific rationalization. To the extent that they still subscribe to the ideology of science, they are then torn between their loyalties to scientific and popular cultures.

Such writers must find strategies for bridging the gap between the two cultures. To do this, however, it is not necessary to study popular culture and attempt to adapt oneself to it as if it were foreign. It is only necessary— but it is also essential—to be attentive to the conflict between the mystic and the technician in oneself. This is by no means an easy task. An author's self-conflict will leave traces in his or her writing, and success will depend in large measure on how well the conflict has been resolved. Let us now look at some examples of scientific writing from this point of view.

We have seen how Feynman expresses his populist side in his colloquial rhetoric without allowing it to interfere with the most fundamental beliefs of scientific culture. Of course, in his *Lectures on Physics*, Feynman was writing for the scientific community rather than a popular one, and his one-sidedness is perhaps to be expected. It becomes more disturbing in his writing for the general public; for example, in his lack of concern for the sociopolitical implications of his work on the hydrogen bomb (Feynman 1985, 118).

Jonathan Miller has probed more deeply into the scientific-popular conflict within himself. It is a great merit in *The Body in Question* (1982) that

Miller shows us clearly how he is torn between his roles as member of both scientific and popular cultures. The very theme of the book exemplifies this conflict, being as it is a delicate reconciliation between Miller's dual careers as neurologist and dramatic producer.

In his popular scientific essays for *Natural History* magazine, Stephen J. Gould (1979, 1980), like Miller, expresses both his role as a scientist and a populist element in his character. But whereas in Miller these two elements are in tension, Gould manages to bring them fruitfully together. He constantly invokes his memories of a childhood full of wonder about the world, and about science in particular—and a child is almost the paradigm for a member of popular culture. Consequently, as a scientist he is willing to examine his discipline from outside itself, in awareness of its history and its significance to the concerns of popular culture.

Finally, in *Silent Spring* (1962), Rachel Carson bases her position on the metaphysics, epistemology, and rhetoric of popular culture. But thanks to her participation in scientific culture, Carson, unlike Maxwell, has no fear of evaluating the credibility of scientific experts and forming her own judgment. The enormous power of her writing comes from blending her authority as a scientist with the moral concerns of popular culture.

Our society desperately needs guidance in the task of trying to understand and control the influence of science. To the extent that scientific writers can free themselves of the ideology of rationalized science (and the examples of Feynman, Miller, Gould, and Carson suggest they can), then scientists themselves can assist with this task by more forcefully expressing the populist part of themselves. They will thereby give a more honest account of the scientific enterprise both as it is conducted within the scientific community and as it relates to the general public.

NOTES

1. The public support for creationism is one example of such a counterattack. See Nelkin (1976). So was the public refusal to accept a ban on saccharin when it was shown to be carcinogenic; for a history of this affair see Wessel (1980). The widespread insistence on the efficacy of Laetrile as a cure for cancer is another example, discussed in Brody (1987), Markle and Petersen (1987), Nelkin (1984), Petersen (1984), Schwartz (1987), and Young (1987).

2. Maxwell unself-consciously uses such terms as *Corinthian column* and *lepidopterans*, decides "it's time for an aesthetically pleasing *objet d'aventure zoologique*," and refers to John Muir and Matisse without explanation; so, enough

education is expected of the readers of *Omni* that they are at least comfortable with these allusions.

3. See Searle (1990), Churchland and Churchland (1990), and Casti (1990). The existence of these semipopular accounts suggests that the challenge of artificial intelligence to human intelligence is growing in importance to the public.

4. For a brief history of the fusion between the mystical and the artisanal strands of science, and the continuing tension between them, see Handlin (1967) and Weil (1967).

5. Each page in Feynman's text is numbered first according to section, then to page.

6. The physicist Jones (1982) compares the influence of science in determining what society calls "useful" to that of the railroads in nineteenth-century America in determining which cities would enjoy success as foci of the new transportation system. Another physicist, Bernstein (1987), makes a similar point.

7. For a critique of this rhetorical device in Bacon and in Sprat's *History of the Royal Society*, see Vickers (1987).

8. For a critique of this customary view of the sociology of science, see Mulkay (1979).

9. A curious coincidence of late medieval or early scientific thought with modern popular culture is mentioned by Gould (1987), who points out the profound resemblances between Thomas Burnet's attempt in the 1680s—early in the Scientific Revolution—at a scientific history of the world, and a contemporary piece of folk sculpture by James Hampton.

10. On how scientific discourse extends into society, see Aronowitz (1988). The same book discusses the extent to which both theoretical Marxism and the actually existing communism of the Soviet Union justify themselves in scientific terms. Nelkin (1987) mentions a few of the antiscientific movements, such as creationism, that have summoned science to bolster their claims. On the intention of creationists to harness scientific language to their arguments, see Moore (1976). He mentions that "all full or voting members of the Creation Research Society must have earned an M.S. or Ph.D. degree in science, or some equivalent." Nelkin (1976) describes the political conflict between the scientific belief system and that of a populist opposition to science, which resembles in many respects the belief system of popular culture as described in this essay.

REFERENCES

Aronowitz, S. A. 1988. *Science as Power: Discourse and Ideology in Modern Society*. Minneapolis: Univ. of Minnesota Press.
Basalla, G. 1976. Pop Science: The Depiction of Science in Popular Culture. In

Science and Its Public. Ed. G. Holton and W. A. Blanpied. Dordrecht, Holland: D. Reidel.

Bazerman, C. 1988. *Shaping Written Knowledge: The Genre and Activity of the Experimental Article in Science.* Madison: Univ. of Wisconsin Press.

Bernstein, H. J. 1987. Introduction. In *New Ways of Knowing: The Sciences, Society and Reconstructive Knowledge.* Ed. M. G. Raskin and H. J. Bernstein. Totowa, N.J.: Rowman and Littlefield.

Brody, B. A. 1987. Quasi Libertarianism and the Laetrile Controversy. In *Scientific Controversies.* Ed. H. T. Englehardt, Jr., and A. L. Caplan. Cambridge: Cambridge Univ. Press.

Carson, R. 1962. *Silent Spring.* Cambridge: Riverside Press.

Casti, J. L. 1990. *Paradigms Lost.* New York: Avon Books.

Churchland, P. M., and P. S. Churchland. 1990. Could a Machine Think? *Scientific American,* January.

Feynman, R. P. 1985. *Surely You're Joking, Mr. Feynman!* New York: Bantam Books.

Feynman, R. P., R. B. Leighton, and M. Sands. 1963. *The Feynman Lectures on Physics.* Vol. 1. Reading, Mass.: Addison-Wesley.

Gould, S. J. 1979. *Ever since Darwin: Reflections in Natural History.* New York: W. W. Norton.

————. 1980. *The Panda's Thumb: More Reflections in Natural History.* New York: W. W. Norton.

———— 1987. *Time's Arrow, Time's Cycle: Myth and Metaphor in the Discovery of Geological Time.* Cambridge: Harvard Univ. Press.

Handlin, O. 1967. Science and Technology in Popular Culture. In *Science and Culture.* Ed. G. Holton. Boston: Beacon Press.

Jones, R. S. 1982. *Physics as Metaphor.* Minneapolis: Univ. of Minnesota Press.

Kuhn, T. S. 1970. *The Structure of Scientific Revolutions.* 2d ed. Chicago: Univ. of Chicago Press.

Latour, B. 1987. *Science in Action.* Cambridge: Harvard Univ. Press.

Levins, R., and R. Lewontin. 1985. *The Dialectical Biologist.* Cambridge: Harvard Univ. Press.

Markle, G. E., and J. C. Petersen. 1987. Resolution of the Laetrile Controversy: Past Attempts and Future Prospects. In *Scientific Controversies.* Ed. H. T. Englehardt, Jr., and A. L. Caplan. Cambridge: Cambridge Univ. Press.

Maxwell, J. 1988. Into the Woods. *Omni,* September.

Miller, J. 1982. *The Body in Question.* New York: Vintage Books.

Moore, J. A. 1976. Creationism in California. In *Science and Its Public.* Ed. G. Holton and W. A. Blanpied. Dordrecht, Holland: D. Reidel.

Mulkay, M. 1979. *Science and the Sociology of Knowledge.* London: George Allen and Unwin.

Nelkin, D. 1976. Science or Scripture: The Politics of Equal Time. In *Science and Its Public*. Ed. G. Holton and W. A. Blanpied. Dordrecht, Holland: D. Reidel.

———. 1984. Science and Technology Policy and the Democratic Process. In *Citizen Participation in Science Policy*. Ed. J. C. Petersen. Amherst: Univ. of Massachusetts Press.

———. 1987. Controversies and the Authority of Science. In *Scientific Controversies*. Ed. H. T. Englehardt, Jr., and A. L. Caplan. Cambridge: Cambridge Univ. Press.

Petersen, J. C. 1984. Citizen Participation in Science Policy. In *Citizen Participation in Science Policy*. Ed. J. C. Petersen. Amherst: Univ. of Massachusetts Press.

Schwartz, R. L. 1987. Judicial Deflection of Scientific Questions: Pushing the Laetrile Controversy toward Medical Closure. In *Scientific Controversies*. Ed. H. T. Englehardt, Jr., and A. L. Caplan. Cambridge: Cambridge Univ. Press.

Searle, J. R. 1990. Is the Brain's Mind a Computer Program? *Scientific American*, January.

Vickers, B. 1987. *English Science from Bacon to Newton*. Cambridge: Cambridge Univ. Press.

Weil, E. 1967. Science in Modern Culture. In *Science and Culture*. Ed. G. Holton. Boston: Beacon Press.

Wessel, M. R. 1980. *Science and Conscience*. New York: Columbia Univ. Press.

White, T. H., trans. and ed. 1954. *The Bestiary: A Book of Beasts*. New York: George Putnam's Sons.

Young, R. S. K. 1987. Federal Regulation of Laetrile. In *Scientific Controversies*. Ed. H. T. Englehardt, Jr., and A. L. Caplan. Cambridge: Cambridge Univ. Press.

Contributors

CHARLES M. ANDERSON is an associate professor of English at the University of Arkansas. He is the author of *Richard Selzer and the Rhetoric of Surgery* (1989) and has published essays on literature and medicine and literature and science.

ANDREW J. ANGYAL is a professor of English at Elon College. He is the author of *Loren Eiseley* (1983) and *Lewis Thomas* (1989).

BRUCE CLARKE is an associate professor of English at Texas Technological University. He edited *The Eighteenth Century: Theory and Interpretation* and coedited, with Wendell Aycock, *The Body and the Text: Comparative Essays in Literature and Medicine* (1990). He has published articles on Anglo-American modernism and the theory of tropes.

MARTIN EGER is an associate professor of physics at CUNY/Staten Island. He is interested in the philosophy of science and education, and hermeneutic approaches to the understanding of natural science.

JEANNE FAHNESTOCK is an associate professor of English at the University of Maryland. She coauthored, with Marie Secor, *A Rhetoric of Argument* (1982) and coedited, with Marie Secor, *Readings in Argument* (1985). She has also

published essays on nineteenth-century British literature.

ROBERT T. KELLEY teaches English at Ursinus College and works as a computer consultant. His research concerns the relations between computer interfaces and narrative form.

MURDO WILLIAM MCRAE is an associate professor of English at Tennessee Technological University. He has published articles on Renaissance literature, literary theory and the philosophy of science, and popular scientific writers.

LOUIS P. MASUR teaches history at CUNY/City College. He is the author of *Rites of Execution: Capital Punishment and the Transformation of American Culture, 1776–1865* (1989) and *Heart-Shaped Leaves: American Writers During the Civil War* (1993).

BARRY PEGG is an assistant professor of humanities at Michigan Technological University. He has published essays on technical writing and the history of printing.

MARY ELLEN PITTS is an associate professor of English at Memphis State University. She has published articles on scientific metaphor and popular scientific writing.

DAVID S. PORUSH is a professor of literature at Rensselaer Polytechnic Institute, where he directs Autopoesis, a team exploring story generation by computer. He is the author of *The Soft Machine: Cybernetic Fiction* (1985) and *Rope Dances* (1979), a collection of short stories. He has written numerous papers about contemporary literature and science and has completed a screenplay, "Mech Head."

DOUG RUSSELL is completing his Ph.D. and tutors in literary theory at the University of Western Australia. He edits a national literary magazine.

DAVID A. STONE is a professor of mathematics at CUNY/Brooklyn College. He has published articles on topology and combinatorial mathematics and has received several National Science Foundation fellowships.

ALLISON BULSTERBAUM WALLACE is an assistant professor of humanities at Unity College of Maine. She is the author of *H. L. Mencken: A Research Guide* (1988) and has published several essays on American literature.

ALAN G. WASSERSTEIN is an associate professor of medicine at the Hospital of the University of Pennsylvania. A specialist in diseases of the kidney, he has published several articles on popular scientific writers.

Index